Volume 3 - The Philippines & Oceania

Author: Abu Hassan Jalil

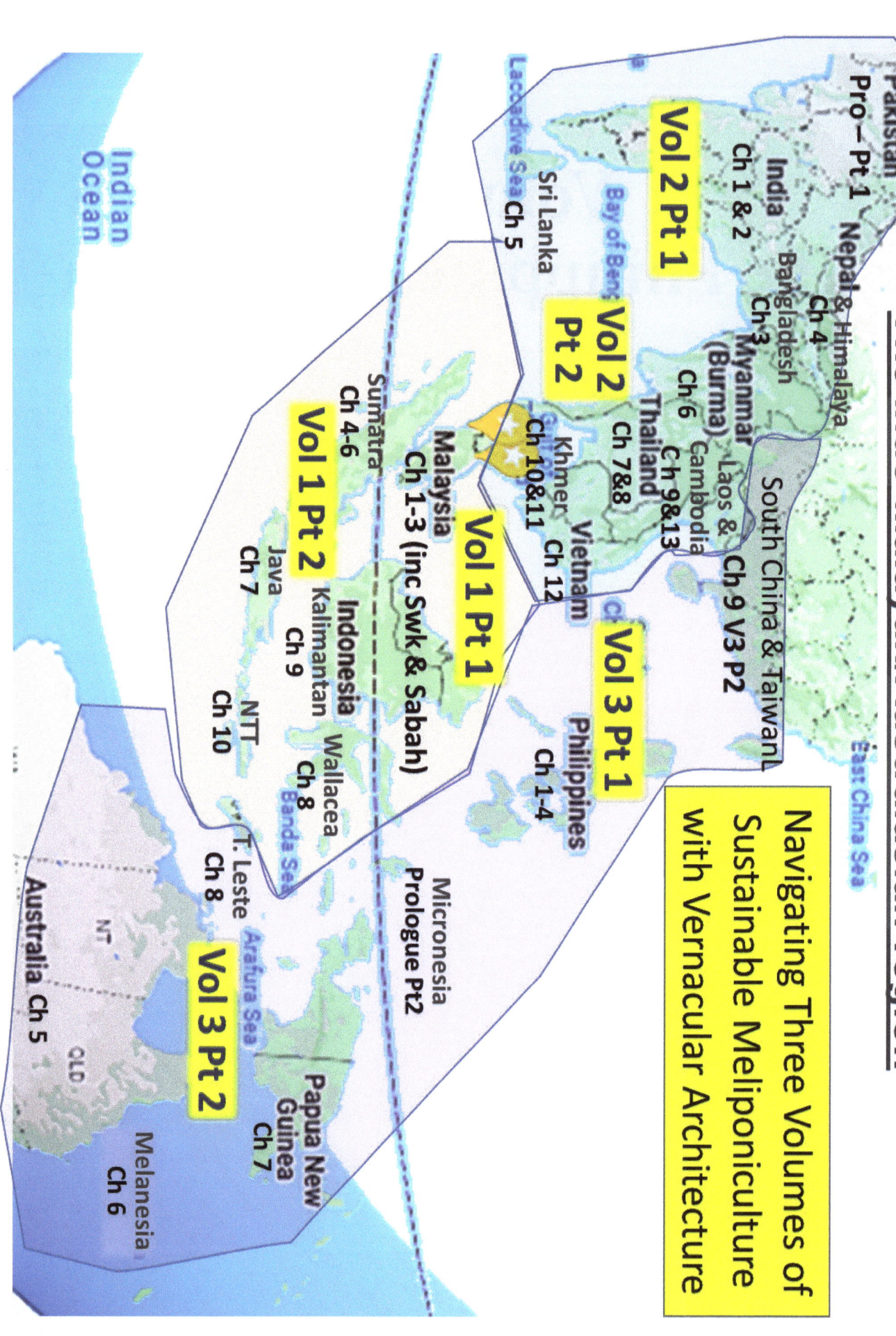

Sustainable Meliponiculture with Vernacular Architecture

Volume 3 - The Philippines & Oceania

Synopsis: This literature proposes new ideas and innovations that consider the ultimate Beescape for Meliponaries, with bee comfort at the top of the list of considerations. Thermal comfort is a subject of grave concern when it comes to conservation in Meliponiculture. Often enough, enthusiastic new beekeepers are not overly worried about heat dissipation, especially during the summers on the fringes of the tropics.

Heat waves in the tropics may leave beekeepers anxious about the stability of the colony's health and population growth. Chances are the existing colonies may see the population dwindling rather than growing at any rate. We also look at earth tremors, if not earthquakes, in the "ring of fire" zone in The Wallacea regions. Flooding threats in Southeast Asia and Typhoons in the Oceanic Philippines are troubling.

Addressing these disastrous events and the potential threats is one of the aims of this book. We have scoured the region and examined how different cultures attempt to mitigate their regions' drastic and extreme weather. Out of these examples and data collection, we provide a gallery of designs on relevant bee housing that may impact such mitigation that may apply to the relevant type of disaster frequently faced in each region.

We also look at bee housing designs in the Australasia region and the peculiarities of extreme weather in Oceania.

Acknowledgements

The author is very grateful to Dr Claus Rasmussen, the Danish entomologist. Department of Agroecology, Aarhus University, Blichers Allé 20, 8830 Tjele, Denmark, for his advice and for sharing the Type Locality of the many Meliponine species in this Indo-Malayan/Australasian region.

We are much obliged to our mentor, Dr. David Ward Roubik of the Smithsonian Tropical Research Institute, Panama, for all his extremely useful and indispensable advice and guidance towards preparing major portions of this book.

We wish to thank Mr. Alexis Dela Cuesta, who organised a webinar as part of Word Bee Day on May 20, 2022, that allowed us to expose the concepts of Sustainable Meliponiculture with pre-colonial Filipino architecture to beekeepers and Jerome Sagcal, who reported these concepts in a Manila Bulletin May 2022 issue (https://mb.com.ph/2022/06/06/beehouse-designs-inspired-by-traditional-philippine-architecture/).

A special mention to Dr. Peejay De Castro of Manuel S. Enverga University Foundation, also known as MSEUF in Enverga Blvd, Ibabang Dupay, Lucena, 4301 Quezon Province, Philippines for his advice and information on some of the indigenous peoples in the Philippines,

We thank Dr. I-Hsin Sung, Assistant Professor, Department of Plant Medicine, College of Agriculture, National Chiayi University, Taiwan, for his invaluable advice and data on Meliponiculture in Taiwan.

Book cover design description:

The Bagobo constitute one of the largest groups among the indigenous peoples of southern Mindanao. Bagobo is the predominant inhabitants of the vast areas extending from the west coast of Davao Gulf to the high reaches of Davao's famous and significant mountain ranges of Mt. Apo or Apo Sandawa to the tribal people.

This scene depicts their unique lifestyle and an impression of their folk architecture should a beekeeper wish to adopt the high-pitch roof for a bee shed or a box hive. The roof provides ample heat dissipation in the hot, humid climate. An ideal design to promote insect/bee tourism.

Their origins can be traced back to the early Malay peoples who came from the surrounding islands of Southeast Asia. Today, their common cultural language and Malay heritage help to keep them connected. They are one of the eight subgroups of the Manobo people of Mindanao.

Contents

Sustainable Meliponiculture with Vernacular Architecture .... - 1 -
Volume 3 - The Philippines & Oceania .... - 1 -
   Acknowledgements .... - 2 -
    Book cover design description: .... - 3 -
Preface .... - 7 -
VOLUME 3, PART 1 .... - 9 -
~ FILIPINO ARCHITECTURE ~ .... - 9 -
    The preamble to Part 1 .... - 10 -
Chapter 1 .... - 12 -
   Traditional Bee Housing, Racks, Shacks, and Sheds .... - 12 -
Chapter 2 .... - 19 -
   Pre-Colonial Architecture in the Philippines .... - 19 -
   Pre-colonial Housing in the Philippines .... - 19 -
      Tribu K'Mindanawan Cultural Village, Davao, Mindanao .... - 29 -
   Moro Traditional Architecture .... - 34 -
Chapter 3 .... - 35 -
   Miscellaneous Philippines Folk Architecture .... - 35 -
   The Maharlika History .... - 35 -
The Maharlika Inspiration .... - 36 -
   Maharlika Grain .... - 37 -
   Bee Shacks in vernacular fashion on Marinduque Island .... - 37 -
   Vernacular architecture in The Zamboanga peninsula .... - 38 -
   The Zamboanga peninsula (https://en.wikipedia.org/wiki/Zamboanga_Peninsula) .... - 39 -
   Vernacular architecture of Palawan .... - 40 -
Chapter 4 .... - 41 -
   Vernacular Bamboo Houses in The Philippines .... - 41 -
   A bee hut in Romblon .... - 42 -
   Sagada House in Tublay, Benguet, Central Luzon .... - 44 -
   Gawahon Eco Park – in Victorias, Occidental Negros .... - 45 -
   Novelty bee housing in Los Pepes, Cavite .... - 45 -
    Ilog Maria Honeybee Farm - Silang, Cavite .... - 46 -
    Norma B & B San Cristóbal in Oriental Mindoro .... - 46 -
   Philippines Honeybees Industries (PHI) .... - 47 -
    Grajos Beefarm in Sorsogon .... - 47 -
    Pia's Bee Farm, Lipa City, Batangas .... - 48 -

    Honey House Bees & Beeks .... - 49 -
        Mikee Balais in Tacloba City, Leyte Island .... - 50 -
    The Kalinga Octagon Bee House .... - 51 -
    Addendum - Vernacular Filipino Construction Terms .... - 52 -

VOLUME 3, PART 2 .... - 55 -

~ OCEANIA OUTLOOK~ .... - 55 -

Prologue to Part 2 .... - 56 -
    Meliponine distribution in Micronesia .... - 57 -
    Republic of Palau .... - 57 -

Chapter 5 .... - 59 -
    An Australian Perspective .... - 59 -

Chapter 6 .... - 65 -
    Melanesia .... - 65 -
    Reviving Fiji's Traditional Architecture .... - 67 -
    The Indigenous Architecture of Vanuatu .... - 68 -

Chapter 7 .... - 69 -
    Vernacular Architecture in Papua New Guinea .... - 69 -
    Bee species with PNG type locality: .... - 70 -
    Vernacular architecture of West Papua .... - 72 -
        The Asmat People .... - 73 -

Chapter 8 .... - 75 -
    Vernacular architecture in Timor Leste and Nusa Tenggara Timor .... - 75 -
    Earth Tremor Resistance Sacred House of Timor Leste. .... - 77 -

Chapter 9 .... - 79 -
    Vernacular Architecture in South China .... - 79 -
    The Origin of Malay Dwelling and the Yunnan Theory .... - 81 -
    Vernacular architecture on Hainan Island, China .... - 82 -
    Lepidotrigona ventralis hoozana in Taiwan: .... - 84 -
        Vernacular Architecture in Taiwan .... - 86 -
        Traditional Houses of Taiwan Tribal People .... - 87 -

References for Stingless Bees in China & Taiwan .... - 88 -

VOLUME 3, PART 3 .... - 90 -

~ ADDRESSING CALAMITIES & VERNACULAR ELEMENTS ~ .... - 90 -
    Introduction to Part 3 .... - 91 -

Chapter 10 .... - 92 -
    Addressing the Earthquake Dilemma .... - 92 -

Most recent Earthquake reports	- 93 -
Traditional Musalaki House	- 94 -
Traditional Nias Island House	- 96 -
Chapter 11	- 97 -
Disaster precautionary measures in flood-prone areas	- 97 -
Chapter 12	- 100 -
Introduction to Vernacular Elements and Architectural Fusion	- 100 -
Chapter 13	- 102 -
Conical roofs and Dome Roofs	- 102 -
Roundhouse and Conical concepts	- 107 -
Novelty conical roof	- 108 -
Beekeeping in the Neotropics	- 108 -
Conical huts of Papua.	- 108 -
Chapter 14	- 111 -
Hanging hives and host containers	- 111 -
Hanging coffins	- 113 -
Mobile Racks for Beehive Housing	- 115 -
Chapter 15	- 116 -
The advent of Green Roofs	- 116 -
Green Roofed Boathouse	- 119 -
Chapter 16	- 121 -
Islamic Structures and Mosques	- 121 -
Chapter 17	- 126 -
Beescape and Wind Effects	- 126 -
Protection From Gales	- 129 -
Beescape Improves Internal Air Circulation	- 129 -
List of Figures	- 131 -
List of Maps	- 136 -
~ APPENDICES ~	- 138 -
Appendix A - Glossary of Roofing Terms and Definitions	- 139 -
Appendix B – List of roof types	- 147 -
Index	- 148 -
Bibliography	- 152 -

Preface

Two native species of stingless bees unique to The Philippines are Tetragonula dapitanensis and Lepidotrigona palvanica, not found anywhere else. They are rather rare and not easily found in their type locality. This book aims to provide information about their existence, consequently calling for greater research on their habitat and nesting peculiarities. All efforts are to conserve these unique bees.

The other stingless bees under culture require housing of more suitable durability considering the frequent natural disasters like typhoons, floods, earth tremors and quakes that spew pyroclastic flows. The predicament the current meliponiculturists face is many, and this book intends to show the remedial steps that can improve and provide sustainable outputs. Among the factors to be considered is thermal comfort for the bee colonies. It is here that vernacular architecture. i.e., one fitting the exact specifications of the environment and using naturally available materials, ... can play a major role in providing a more lasting surrounding and home for the bees.

Regardless of faith or ethnic culture, religious stricture tends to be more stable and durable. They are structures generally built by the community for expected use by many generations. This literature examines Christian churches, Muslim mosques, paganistic faiths, etc. They all provide vernacular structures that are used and relied upon. These communities usually blend the natural elements with available materials and climatic conditions.

Reviewing the Qur'anic verse ٦٨ 16:68, "*Make your homes in the mountains, the trees, and in what people construct*", is the motivating factor that edges this author to delve into **what people construct** that inspires the bees to make their homes. Undoubtedly, what people construct must be indigenous dwelling structures of folk architecture. Therefore, as a beekeeper, one must digest this philosophy and appreciate bees' visit to one's home is not an intrusion but a blessing. Hence the theme "Sustainable Meliponiculture with Vernacular Architecture".

Even though I was motivated by certain verses in the Holy Quran, the term "what people construct" does not mention the religious identity of the structure. I remember once in the village of Tunga, in Leyte Island, Visayas in the Philippines, I was asked to view a peculiar nest entrance of stingless bees in a Chapel wall. The cerumen was off-white, and the wall was whitewashed. The whitewash did not contaminate the nest entrance; cerumen usually turns white when damp and cold. We realize that bees are not overly critical about utilising the structure they harbour in. Be it mosques, chapels/churches, synagogues, temples, pagan structures or even skull collectors' or headhunters' dwellings, bees may find a way to nest alongside and within.

They look for or respond to, above all, a) sustainable thermal comfort, b) a haven from detrimental winds (turbulent eddies and bad venturi effects), c) appropriate cavity size, d) comfortable humidity, e) protection from continuous downpours, etc. The ideal nest can be arboreal or subterranean if the conditions are not erratic and inconsistent.

In my travels in the Philippines, specifically in the Visayas, where traditionally built ancestral homes still exist, many had numerous stingless bee colonies. Chris Starr and Shoichi Sakagami, in a publication from 1987 (and reviewed by D. W. Roubik then!), depict a house of a native family in Tabuc-Tubig, Dumaguete, Negros Oriental. It had bee-inhabited bamboo beams of that traditional house. There was a total of 48 unbroken cavities in the beams. They were all occupied by colonies believed to be *Tetragonula fuscobalteata* or *T. sapiens*.

Henceforth, Part 1 starts with how vernacular architecture relates to traditional bee housing, racks, shacks, and sheds in The ASEAN region. How one can not only boost tourism but also create innovative ideas for bee housing with examples from unique Island cultures and long-lost native structures are themes that are used throughout. We go forth with Insect and/or Bee Tourism in mind and a penchant for reviving Pre-Colonial Architecture in the Philippines.

Pre-colonial Housing in the Philippines will explore traditional homes in the Cordillera region of Luzon, the many islands of the Visayas with the wide array of tribal cultures of Mindanao. This section of Part 4 explores as many as 26 tribal houses and describes each. In some, we will see the stages from the Pre-colonial structure to the Spanish era to the current organic village dwellings. A virtual visit is made to the "Tribu K'Mindanawan Cultural Village" in Davao, Mindanao.

There are examples from a bee hut in Romblon; a Sagada House in Tublay, Benguet, Central Luzon; village structures turned bee gallery in Lipa, Batangas, South Luzon; some gazebos in Bulusan Sorsogon, Bicol; a *Bahay Kubo* or Nipa Hut in Gawahon Eco park in Negros Island; a new spread for visitors in San Cristóbal in Oriental Mindoro; a bee sanctuary in the making in Leon, Iloilo, Pany Island. These examples show that structures made by man with appropriate and effective passive ventilation seem equally appropriate for the native bees.

As a complementary note to vernacular structures in Oceania, besides the Solomon Islands and Fiji, we also provide an outlook for the Australian Bee housing. The Australian beekeepers are constantly innovating and improving their hive boxes, which is desirable that the Indomalayan Meliponiculturists should emulate.

VOLUME 3, PART 1
~ FILIPINO ARCHITECTURE ~

The preamble to Part 1

The Philippines is inhabited by more than 182 ethnolinguistic groups, many of which are classified as "Indigenous Peoples" under the country's Indigenous Peoples' Rights Act of 1997. Traditionally, Muslim peoples from the southernmost island group of Mindanao are usually categorized together as Moro peoples, whether they are classified as Indigenous peoples or not. About 142 are classified as non-Muslim Indigenous People groups, and about 19 ethnolinguistic groups are classified as neither Indigenous nor Moro.

The Muslim-majority ethnic groups and ethnolinguistic groups of Mindanao, Sulu, and Palawan are collectively called the Moro people, a broad category that includes some indigenous and some non-indigenous people groups. With over 5 million people, they comprise about 5% of the country's total population. The Spanish called them Moros after the Moors, despite no resemblance or cultural ties to them apart from their religion.

We then look at these indigenous people's ancestral homes. They have feral colonies of native bees in their walls, posts, door jambs and foundations. To the Moro groups, this phenomenon is natural to bees. They term it "Silatur an Nahl", meaning friendship with the bees.

Quoting a passage revered in the Holy Quran since more than 1400 years ago.

(https://quran.com/16) An-Nahl (The Bee)

وَأَوْحَىٰ رَبُّكَ إِلَى ٱلنَّحْلِ أَنِ ٱتَّخِذِى مِنَ ٱلْجِبَالِ بُيُوتًا وَمِنَ ٱلشَّجَرِ وَمِمَّا يَعْرِشُونَ ٦٨ 16:68

And your Lord inspired the bees: "Make ˹your˺ homes in the mountains, the trees, and in what people construct",

16:69 ٦٩

ثُمَّ كُلِى مِن كُلِّ ٱلثَّمَرَٰتِ فَٱسْلُكِى سُبُلَ رَبِّكِ ذُلُلًا ۚ يَخْرُجُ مِنۢ بُطُونِهَا شَرَابٌ مُّخْتَلِفٌ أَلْوَٰنُهُۥ فِيهِ شِفَآءٌ لِّلنَّاسِ ۗ إِنَّ فِى ذَٰلِكَ لَءَايَةً لِّقَوْمٍ يَتَفَكَّرُونَ

and feed on ˹the flower of˺ any fruit ˹you please˺ and follow the ways your Lord has made easy for you." From their bellies comes forth liquid of varying colours, in which there is healing for people. Surely this is a sign for those who reflect.

About 142 of the Philippines' Indigenous People groups are not classified as Moro peoples. Some of them are grouped due to their strongly associated with a shared geographic area, although the ethnic groups themselves do not always welcome these broad categorizations. For example, the Cordillera Mountain Range indigenous peoples in northern Luzon are often referred to using the exonym "Igorot people" or, more recently, as the Cordilleran peoples. Meanwhile, the non-Moro peoples of Mindanao are collectively referred to as the *Lumads*, a collective autonym conceived in 1986 to distinguish them from their indigenous Moro neighbours.

Introduction to Part 1

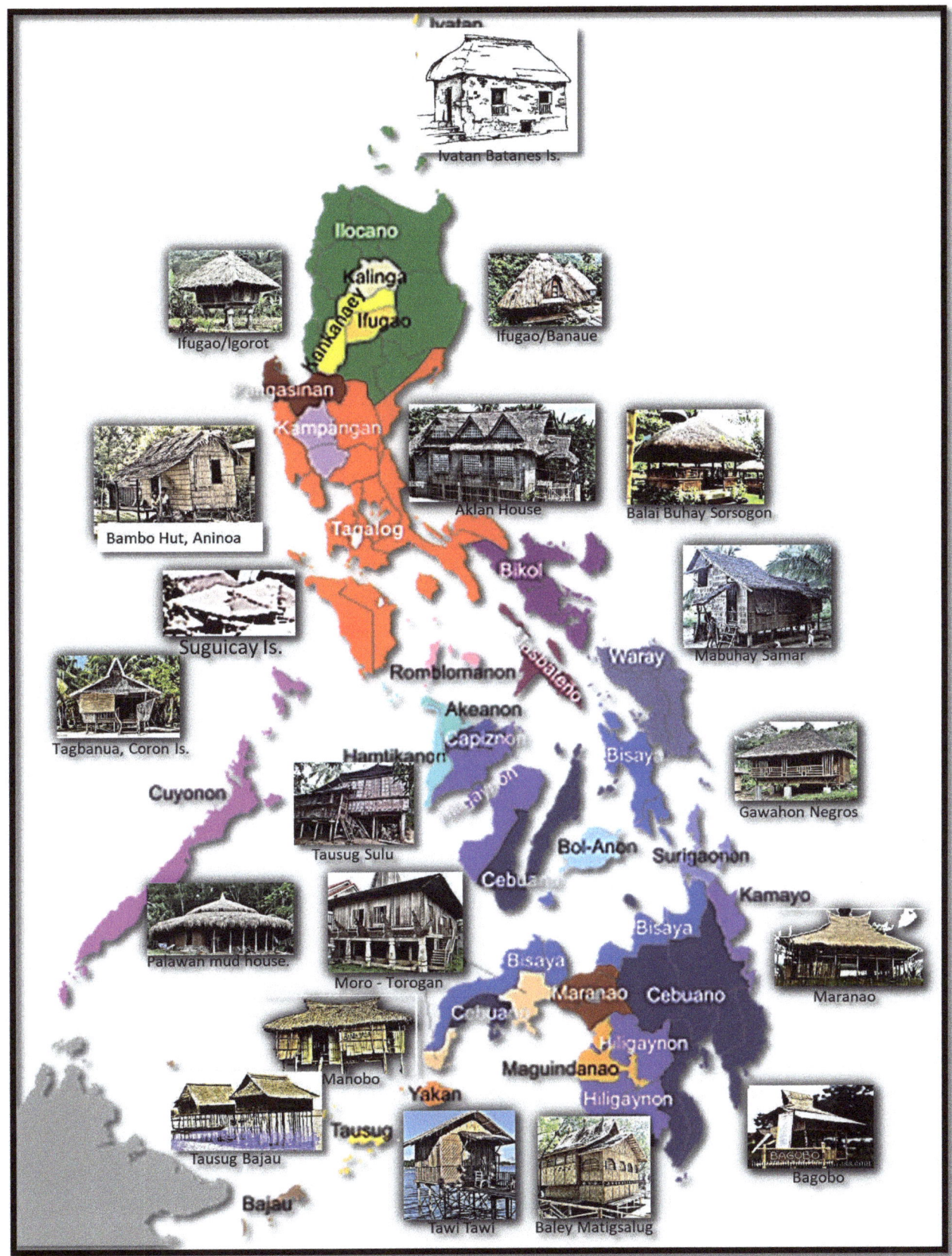

Figure 1 Vernacular architecture in the Philippines.

The varied vernacular architecture of different islands of the Philippines shows the diversity among the indigenous structures and dwellings.

Chapter 1

Traditional Bee Housing, Racks, Shacks, and Sheds

Short story: Once, a tortoise named JOO happened to be in a village where Tortoise soup was a much sought-after delicacy. Stuck in that predicament, he never put his head out for fear of it being chopped. From inside JOO's shell, he could see villagers building odd structures, but he thought they must be inspired by his house/shell.

Figure 2 Almost completely enclosed roofing structures for protection against extreme weather.

JOO thought the structures must be a sacrificial dorm for him because they used straws, palm leaves and dried grasses. These are occasionally his dietary requirements. Not sticking his head out, he could only watch and deduce from views inside his shell. Not knowing that the houses were built to withstand strong winds because the village was in a typhoon-prone area.

He also did not know that globally, in the Tropics, straws, palm leaves and dried grasses are used for structures traditionally.

Figure 3 Example Vernacular in Global Meliponaries.

I borrowed these from some issues of the 'Bees *for* Development journal.

Figure 4 Aguaruna people building their houses in Bajo Cachiaco, Peru.

An Aguaruna man and his family and friends are adding palm leaves to the roof of his house in Bajo Cachiaco, Peru. It appears that they make a round-ended house with thatch and natural materials. (Jernigan Kevin 05/2006 on ethnobiology in Peru).

JOO kept his head safe in his shell and NEVER got anywhere. Tortoises can only move if they stick their neck out. This thought sets me out to pursue something I observed on Panay Island. We found an ancestral house in Awis municipality with 109 colonies in the woven bamboo walls. I started looking for more traditional homes as I travelled throughout the Philippines.

Another short story (true): Before the global COVID-19 pandemic 'lockdown', I travelled with Dr. Roubik and Ms. Nora Peña to Palawan in the Philippines. We were coming out of Aborlan, and something caught my eye. I told the driver (Jun Tabinga), "Stop, stop!" and Dr. David Roubik asked, "WHY". I said I saw a unique house and wanted to snap a picture. He looked angrily (as it appeared to me then) at me and frowned. I was adamant (I felt... a man has got to do what a man must do). I went out, walked a short distance into a side road and took a couple of pics. I walked back prepared for an excuse. Climbed into the car, apologized, and said I could not resist recording the image. He said, "It's ok, only that there is no such thing as VERY unique. Unique is unique; 'very' in this instance is just a redundant superlative (or something to that effect). Hah! Even in conversation, he edits me. We all (including the driver) just laughed and went along our way back to

Figure 5 Typical Tausug Architecture influence on this

Puerto Princesa. Note to self: If you want to learn, then listen well. So, this is the photo I took (Figure 6). Now, isn't that unique compared with others in the Philippines? I photographed some traditional huts and houses during my travels in The Philippines (Figure 7), some structures used for bee sheds, and some were devastated by Typhoons.

Figure 6 The uniquely constructed house in Palawan

Figure 7 Leo Grajo's new setup devastated after Typhoon

They must develop new and ingenious innovations to withstand Typhoons for their racks and shacks.
Some ideas in Wallacea may help arrest structural wind instability in strong breezes and gale-prone areas.

Figure 8 Traditional Huts and Houses I photographed during travels in The Philippines.

These are house models in the Philippines:

1. Saltbox roof - features two slopes of varying length, one much longer than the other. Because of its sloped structure, water runs off easily on this roof, making it ideal for areas that receive heavy rains.

2. Conical - Also known as a cone roof, witch's hat, or turret roof, a conical roof is round on a flat pane and rises to a point, forming a cone shape. Like other common roofing styles such as gable and hip roofs, conical roofs also have roof rafters and support columns, although, most of the time, they are shaped unusually and cut at different angles to match the cone form. Equal protection from any direction of the wind.

3. Jerkinhead roofs - offer more attic space and greater wind stability. This roofing type has a complicated design, making it costlier than a standard gable or hip roof.

4. Hexagonal roofs are essentially roofs with six sides that slope downward. There are two structures I saw on Filipino bee farms that excited me. One is the Hexagon Beehouse in Grajo's Bee Farm in Sorsogon.

Figure 9 Hexagon Beehouse in Grajo's Bee Farm in Sorsogon.

Figure 10 Modifications on the Hexagon Bee house in Grajo's bee farm, Sorsogon

Pavilions, cabanas, and gazebos are some of the most common structures that use hexagonal roofs. They are mainly used to improve a structure's aesthetic appeal rather than utility. The one in Grajo's has an elevated top raised to allow air circulation. Leo Grajo has since modernised the structure, which now looks more solid.

The Philippines is a 'multi-island' nation, and it is no surprise that its inhabitants favour beach huts, cabanas, gazebos, and pavilions as rest huts in their gardens and farms. Multi-sided structures allow multi-directional take-off for the bee's foraging flight. Some designs may provide suitable features for a Bee shed or Bee racks housing shelter.

Figure 11 Gazebos, Cabanas and Pavilions as Bee sheds or Bee rack shelters.

It is a matter of getting the appropriate shape and roof that reflects one's ethnic background or tribal or clan heritage. These structures can be utilized for both human and bee comfort.

5. Tiered Pyramid Roof - The other structure (contd. From No.4) is in Occidental Mindoro at Danizon farm. The structure is his house, which he says is Balinese-inspired. This structure formed the basis of the following rack design (Figure 14).

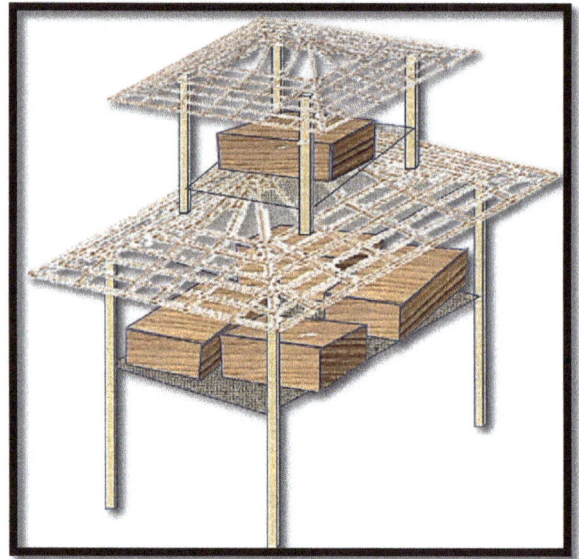
Figure 13 Balinese architecture inspired Bee Rack

Figure 12 Lakeside Two-Tiered Bee Pavilion

The possibilities are endless, and mass concrete may strengthen and anchor the pillars. This bee box hives rack with the house and other structures keeping in line with the whole concept will make a great attraction for visitors.

Figure 14 Danny Hizon's house in Sabluyan, Occ. Mindoro; Right: A Balinese structure that inspired Danny Hizon

Chapter 2

Pre-Colonial Architecture in the Philippines

Pre-colonial architecture started with the influences of the Malays. Designs during this period vary by region, but common features include a steep roof over a one-or-two-room living area raised on posts or stilts one to two meters above the ground or over shallow water. However, the earliest shelters of the Filipinos were probably not built by them.

Pre-colonial Housing in the Philippines

Referred[1] to as FOLK Architecture, INDIGENOUS Architecture, PRIMITIVE Architecture, and Vernacular (native) architecture found among the ethnolinguistic communities in the country. All are correct. This means they draw inspiration and demand mainly from our environment, specifically the climate, terrain, vegetation, and fauna. It is also based on available materials and responds to communal and social needs to be safe from hostile and marauding tribes and to interact with fellow human beings. Let's see some pre-colonial houses from Luzon to Visayas & Mindanao.

Luzon

The **Ivatan**[2] houses are not the typical houses you can find in the Philippines. The Ivatan houses are made of metre-thick limestone and coral walls as well as cogon grass roofs, and it is sturdy enough to withstand strong winds. Typhoon and strong gale-resistant, the rubble-walled Ivatan houses of Batanes

Figure 15 The Ivatan House a.k.a. Kalinga House in Batanes Island, Philippines. A Sinadumparan Ivatan house, one of the oldest structures in the Batanes Islands. The house is made of limestone and coral and its roofing is of cogon grass.

Island in the Northern part of the Philippines. Batanes Island, isolated to the North, is known to face

---

[1] Glendz Madoro https://www.academia.edu/39626218/PRECOLONIAL_HOUSING_IN_THE_PHILIPPINES
[2] https://en.wikipedia.org/wiki/Ivatan_people

extreme weather conditions, and the houses withstanding the extremes are testaments to its construction durability.

Before the Spaniards arrived in the Philippines, Ivatans built their houses from cogon grass. These homes were small, well-situated, and designed to protect against strong winds. The Spaniards introduced large-scale lime production to the Ivatan to construct their now-famous stone houses. The basic cogon grass is still preserved as roofs of their houses, thickly constructed to withstand strong winds. Meter-thick limestone walls are designed to protect against the harsh Batanes environment, which is known as a terminal passage of typhoons in the Philippines.

Figure 16 Inspired by the Ivatan limestone rubble wall house.

The **Ifugao** houses were usually similar in architectural designs but differed in decorative details depending on the tribes. Their houses were harmoniously located with the contour of the rice terraces. The one-room house of the Ifugao is commonly known to them as fale[3].

Figure 17 Ifugao house or Igorot Hut

---

[3] Samoan house with open sides and a thatched roof.

The **Kankanay** House is a vernacular house in which the roof provides vents over the storage platform over the dapugan. In contrast to the cluster above, this house, with its run fence, prefers isolation.

Figure 18 The Kankanay House

Ibaloi house. Ibaloi is derived from 'i-', a prefix signifying "pertaining to" and badoy or house, together meaning "people who live in houses". The Ibaloi are indigenous peoples collectively known as Igorot (igudut, "hill-dwellers") who live in the cordillera central of Luzón.

A design is unique to the Isneg, who are boat

Figure 19 Ibaloi house.

builders. The **Isneg** house, built slightly above the ground, is also a one-room dwelling like the fale but has lighter & is bigger. Its bamboo-layered gabled roof ensembles an inverted boat.

Figure 20 The Isneg house.

Octagonal House of the **Kalinga**. ... The southern Kalinga cultivate rice on terraces and in swidden, and they have settlement patterns with up to two hundred houses with patches of small villages. The northern Kalinga, engaging in swidden[4] farming, settled in dispersed hamlets of six to thirty houses.

Figure 21 Gilitob - Octagonal House of the Kalinga.

The pre-colonial era saw the Use of available materials like stacking sapling trunks to cordon off domesticated animals below the house. Thick tree branches as wall posts and walls may be tree bark.

The newer styles show timber slat walls, and the cordoned confinement is improved with woven bamboo. The posts are made of bunched bamboo, and later, lumber is used as and when materials are available. Cogon grass still covers the roofs but with different configurations. (See Figure 22)

Kalinga Culture

There are four types of Kalinga Traditional Houses (Beyoy):
1) Gilitob - A wooden Octagonal shaped house of the Kalinga nobles called "Kachangyan"; Additional info: The "Pangat" (Tribe leader) came from the Kachangyan families.
2) Piletong - Wooden Square shaped house of wealthy families (Byaklang);
3) Bilulilaw - Bamboo panel house of marginal families (Kapus); and
4) Takkubi - Shanty dwelling of the impoverished.

---

[4] Swidden farming, also known as shifting cultivation or milpa in Latin America, is conventionally defined as "an agricultural system in which temporary clearings are cropped for fewer years than they are allowed to remain fallow" (Sanchez, 1976).

Kalinga Architecture Trivia[5]:

Early Kalinga Houses were built without using a single nail. They use "*peyong*" instead (wood joints/fasteners or dowels), which is more durable and can last longer since it allows the wood to move with seasonal shrinkage and expansion.

Such ancient architecture is also perfect for a tropical country like the Philippines. It allows whole-house ventilation because of the elevated posts, bamboo mats called "*littagon*" as floorings, and front and back outlets.

The Tinguian[6] and Ilocano... They left their ancient home as a unit before the Hindu domination of Java and Sumatra. Upon the arrival of Salcedo, the greater portion of the coastal people accepted the rule of Spain and the Christian religion, while the more conservative element retired to the interior and merged with the mountain people.

To the Spaniards, the Christianized natives became

Figure 22 The Tinguian House

known as Ilocano, while the people of the mountain valleys were called Tinguian, or mountain dwellers. The Tinguians are pagan Philippine people who inhabit chiefly the mountain province of Abra in northwest Luzon.

Figure 23 The traditional Bontoc house

The pre-Christian **Bontoc** belief system centres on a hierarchy of spirits, the highest being a supreme deity called Lumawig. The traditional Bontok house was made of wood or cogon grass.

---

[5] Source:
National Living Treasure/Gawad Manlilikha ng Bayan, Apo Alonso B Saclag, Sr.
Model houses on-site at the Awichon Cultural Village, Lubuagan, Kalinga and the Nayong Pilipino, Clark Field, Mabalacat, Pampanga
[6] Source: https://www.slideshare.net/viganIlocostelenovela/research-on-the-tinguian-44256922

The Aeta (Ayta /ˈaɪtə/ EYE-tə), Agta, or Dumagat are collective terms for several Filipino indigenous peoples who live in various parts of the island of Luzon in the Philippines. **Aeta** is pronounced "Eye Ta." They are indigenous, and their ancestors were the Aborigines from Australia. There was a consensus from anthropologists that they migrated from the island of Borneo about 30 thousand years ago using a land bridge partially covered by water 5,000 years ago.

Figure 24 The Aeta's house

The Aeta's house is made of bamboo and cogon grass. Most can be found in the mountains.

Majukayongs[7] humbly accept their role as heirs and successors of the Maducayan Land. In its geographic meaning, it refers to those who reside or reside in the jungle of Maducayan, Saliok and some other parts of Tanudan, Paracelis and Natonin.

The Visayas

The **Mangyan** House walls are made of the bark of trees and constructed about a meter or less above the floor - this opening allows occupants to observe the exterior without having seen it from the outside.

Figure 25 The Mangyan House

Figure 26 The Tagbanwa House

The **Tagbanwa** House comprises bamboo & wood for a strong frame, *anahaw*[8] leaves for the roof and walls, and bamboo slats for the flooring.

See also "Vernacular Architecture of Palawan" p. 38 and Figure 64.

---

[7] https://www.facebook.com/maducayan/
[8] Fan Palm leaves.

Mindanao

Torogan, a "resting place" or "sleeping place", is a traditional house built by the **Maranao** people of Lanao, Mindanao, Philippines. The Maranao people (Maranao: ['mәranaw]; Filipino: Maranaw), also spelt Meranao, Maranaw, and Mëranaw, is the term used by the Philippine government to refer to the southern tribe who are the "people of the lake", a predominantly Muslim Lanao province region of the Philippine Islands of Mindanao. A *torogan* was a symbol of high social status. Such a residence was once home to a sultan or Datu in the Maranao[9] community.

Figure 29 A 300 yr old Maranao House. Inset: Top right: Human scale for the size of the tree trunk post; Bottom right: closeup of stone castors beneath the Tree trunk pillars.

The house is in Lanao del Sur (Lake Lanao, South Lanao Subdistrict in Mindanao). The Maranao people are called the Danao people [Malay – Danau for Lake] because they are domiciled in Lake Lanao) The Mindanao region (where Zamboanga is located) gets its name from them. Incidentally, the large

Figure 28 Left: Sarimanok or (Papanoka Mra) is a legendary bird of the Maranao; Middle: A Maranao house preserved amidst modern buildings in the background at Marawi City, Lanao del Sur, Mindanao, Philippines; Right: Model of Torogan house of the Maranao people at Cockington Green Gardens

southern island of the Philippines is named Mindanao Island. They are still descendants of the Sultan of Lanao.

Figure 27 Maranao Carvings

---

[9] https://en.wikipedia.org/wiki/Maranao_people

Image source: 1. Sarimanok (*Papanoka Mra*) is a legendary bird of the Maranao. The carved wings jutting at the posts are akin (presumably) to the wings of the Sarimanok.

Figure 30 Inspired by the Traditional Maranao house with rock castors beneath tree trunk posts.

2. Model of Torogan house of the Maranao people at Cockington Green Gardens is a park of miniatures in Nicholls, Australian Capital Territory, Canberra. The following images illustrate variations and modifications to the Maranao House Design.

Figure 32 Maranao Royal House with Princess' Chamber

Figure 31 The Classic Maranao Torogan

Figure 34 Modified Maranao Torogan

Figure 33 Maranao Bee Gallery

The **Yakan** house is usually scattered among the fields, and it isn't easy to see where one settlement ends and the next begins. The inhabitants of a settlement may or may not be of the same clan. The Yakan have a traditional horse culture. They are renowned for their weaving traditions. The Yakan people are among the major indigenous Filipino ethnolinguistic groups in the Sulu Archipelago.

Figure 35 The Yakan house

The **Tausug** house is a native tribe in Jolo in the Sulu archipelago of the Philippines. • The name Tausug is derived from two words: "*tau*", meaning man, and "*sug*"or "*suluk*", meaning current, thus making Tausug "people of the current".

Figure 36 A Tausug House with 109 tiny SB colonies in the bamboo walls.

Figure 37 The Bajao House

Figure 38 The Mandaya House

The **Bajao** House. The Samal houses are grouped in villages and are connected by bridges and catwalks. They build their houses on stilts over the water, along the shore, or farther out.

The **Mandaya** house is made of carefully selected bamboo flattened into slats and held together by horizontal bamboo strips or a rattan. The ascent to this single room with a small kitchen area is through a removable single-notched tree trunk. Traditionally, its elevated floor line served as one of the safety measures against attacks from other ethnic groups in the periphery of Davao Oriental. These warring conditions made the *Bagani* or warrior class a high and most coveted social ranking.

T'boli tribe

Figure 39 The Gono Taug'na House of the T'boli Tribe

The **T'boli** tribe dwells on elevated terrains, particularly around Lake Sebu, Lake Selutan, and Lake Lahit in South Cotabato in Southwestern Mindanao. The communities generally comprise several kin-oriented clusters with extended families common in households.

Their traditional house, *Gono*, is built mostly of bamboo, cogon, and *uway* (rattan). The house is made of woven bamboo strips, elevated three feet or more above the ground and supported by bamboo poles, while the floor is made of bamboo lath. The roof is made of dried cogon grass intricately tied with bamboo laths. *Uway* or rattan strips tie the components together. The size and complexity of the house are relative to the economic and social class of the dweller. The **Datu**, the head of the community, usually has the biggest *Gono*.

Each house has a different shape and style but a unique design. Most ethnic houses conform to a general pattern, like steep thatched roofs to facilitate drainage, elevated on posts or stilts to temper the earth's dampness and humidity, slated flooring to let in cool air from below, tight-fitting solid planks to help keep upland huts warm. The types of dwellings are highland, lowland dwellings & coastal, and river.

Tribu K'Mindanawan Cultural Village, Davao, Mindanao

(excerpts from http://davaocitybybattad.blogspot.com/2012/11/tribu-kmindanawan-cultural-village.html)

Tribu K'Mindanawan is a cultural village in Davao City that showcases the richness and diversity of the various cultures and colourful heritage of the different indigenous tribes of Mindanao. The village consists of the cultural communities and tribes of the Mangguangan, Mansaka, Ata Manobo, Ata Paquibato, Obo Manobo, Bagobo Tagabawa, T'boli, B'laan, and Mandaya.

Mangguangan

Figure 40 Tog'gan House of the Mangguangan Tribe

The *Mangguangan* tribe is a close kin of the *Dibabawon* tribe that dominates New Corella's highland regions in the Davao del Norte province. Their traditional house, called the *Tog'gan*, is made mostly of a *lawaan*[10] bark - a light yellow to reddish-brown or brown wood, while the roof is made of rattan leaves lashed on bamboo laths supported by a set of *kawit* or *sungag* (finials) made of *bagakay* bamboo attached at the gable ends of the roof which is believed to ward off evil spirits.

The *Tog'gan* is usually elevated three feet above ground, supported by sturdy round timber posts, and the walls are only about shoulder height. A small halaran altar is at its door, also believed to protect the household from evil spirits. An outdoor altar called *inang* is placed in front of the house during a blessing.

Mansaka

Early Mansaka houses were built on treetops or bamboo groves as a precaution against surprise attacks and raids. The *Mansaka* is the most prominent tribe that lives in dispersed settlements in the fertile valleys and hills of Compostella Valley Province, whose economic and political life is largely guided by tribal elders known as *Matikadong*. The *Baylan,* or

Figure 41 Uyaanan House of the Mansaka Tribe

---

[10] *Xanthostemon verdugonianus* - Wikipedia https://en.wikipedia.org › wiki › Xanthostemon_verdug... The bark is slate white in colour and has a peeling appearance.

the village shamans, perform rituals for the tribe, while the *Bagani* or tribal warriors, protect the rights and lives of the people.

Their traditional house, the *Uyaanan,* is typically a single-room structure built on top of a tree some twelve feet above the ground, supported by sturdy timber posts. The house is usually made of bamboo, while the roof is made of *sasal* or bamboo shingles. A single log carved with notches or footholds serves as a ladder to the house. Traditionally, the front of the house, called *papaudan*, faces the morning sun.

The Manobo house, Agusan. Manobo means "people" or "person"; alternate names include Manuvu and Minuvu. ... The Manobo usually build their villages near small bodies of water or forest clearings, although they also opt for hillsides, rivers, valleys, and plateaus.

Ata Manobo

Figure 43 Binotok House of the Ata Manobo Tribe

The **Ata Manobo** is one of the three tribes that live in the hinterlands of Talaingod in the province of Davao del Norte, who traditionally practice slash-and-burn cultivation. Their traditional house, the Binotok, is typically constructed on hills or ridges adjacent to the fields and is usually far from neighbouring homes. It is usually square and is constructed low, sometimes perched on top of a tree's stump, with half-walls made of *lawaan* or *langilan* bark[11], while the roof is made of thick layers of **cogon**. The floor is made of

Figure 42 Left: Gantangan is an old measurement of selling rice in the absence of a weighing scale. Contributed by Alberto V. Bartolata; Right: . Ata Paquibato - Gantangan Palabain ref: https://m.facebook.com/watch/?v=1363139273747172&_rdr

---

[11] BOTANICAL NAME: *Cananga odorata* Hook. f. et Thorns (Annonaceae)

bamboo laths lashed on round timber joists, supported by sturdy round timber poles. The house is raised high above the ground with a single log ladder usually drawn up at night. The space below is used mainly for storage, where an outdoor altar called the *ankaw* can also be found. In some instances, wooden rat guard discs are added to the posts of the house.

Ata Paquibato

The ***Ata Paquibato*** is a mixed 'Negrito' tribe that dwells in the upland district of Paquibato in Davao City, led by a local chieftain called *Datu*. Their ritual leader, known as *Temanerun,* is believed to possess a spirit guide commonly known as *Abyan*.

Gantangan (Figure 44) is an old measurement for selling rice without a weighing scale. It has the same meaning in Malay and Indonesian languages as well. The Gantangan Image is from Alberto V.Bartolata.

Gantangan is used to measure a ganta of rice, beans other farm products. That house is an extension of a rice barn. "gantangan" refers to a measuring box used to measure a ganta equivalent to 2 and 1/4 kilos.

The word "palabain" is in a Hiligaynon or Visayas dialect, and the Filipino term is "pahabain", which means "extend" or "elongate" in English—according to Nora Alanza Peña.

The illustration (Figure 43) above uses the term for an indigenous dwelling model and is an extension of a grain store or barn, now modified as a bee house to accommodate hanging hives because of its balcony.

Unlike the *Lumads*, the traditional house, called *Binanwa Baluy*, is built perpendicular to the east, or the *sidlakan,* because it is popularly believed that facing the morning sun will bring misfortune upon the dwellers. The tribe believes in spirits borne from nature, and a house cannot be built without the *Temanerun* performing the rituals of *panubad-tubad* and the *pamalas* that seek guidance and permission from the spirits, where an offering is placed in a *tambalan* consisting of at least three altars built in front of the house.

The house is usually elevated on stilts made of sturdy round timber posts and, in some instances, is perched on top of a tree stump as a defence against *pangayaw* or enemy raids. The house is mostly constructed of *lawaan* bark[12] lashed on a bamboo lath framing, while the floor is made of bamboo laths. The roof is a bowed ridge beam of *cogon* grass braided on bamboo purlins where carved figures

---

[12] *Xanthostemon verdugonianus (Myrtaceae)*, commonly known as mangkono or Philippine ironwood

of a horse usually adorn the gable ends of the roof. Small windows on the side and rear walls serve as lookouts against potential spear-wielding attackers. The house can only be accessed through a single log ladder drawn up at night.

Obo Manobo

Figure 44 Bakag Farmhouse of the Obo Manobo Tribe

While the **Obo Manobo** tribe are generally upland farmers, they also hunt and fish. The tribe dwells on steep slopes or ridges in the highland district of Marilog in Davao City, whose settlements are kin-oriented.

Their traditional house, *Bakag*, is an airy single-room structure perched on round timber posts elevated above the ground and can comfortably accommodate two to three families. It is usually constructed of *salahiya* or woven bamboo strips, rafters, beams and roof frames of assorted round timbers, tied together with *Uway* or rattan strips, while the floor is made of bamboo laths. The roof is made of *cogon* and rattan leaves lashed on bamboo lath rafters. Built beside the house, about four to five feet above the ground, is a small vault called *lapong* where the rice supply and the seeds for the next planting are stored.

Bagobo Tagabawa

The **Bagobo Tagabaw**a tribe is one of the three subgroups of the **Bagobo** tribe that predominantly inhabits the vast tracks of land extending from the west coast of Davao Gulf up to the fertile hills and valleys at the foot of Mount Apo.

Figure 45 The traditional house of the Bagobo Tagabawa of

Their houses are typically built on top of hills and are scattered near the field. The traditional house of the **Bagobo Tagabawa** of Tibolo, Sta. Cruz, called *Bale*, is made mostly of bamboo. The walls and distinctively steep roofs are all *sinasa na kawayan* or flattened bamboo. The floor and eaves are made of *linapakan* or bamboo laths[13], while the posts and beams are round timber.

---

[13] Thin flat strips of wood or bamboo, especially one of a series forming a foundation for the plaster of a wall

B'Laan

The ***B'Laan*** tribe dwells on the hills behind the western coast of Davao Gulf, extending up north to the Bagobo territory and westward into the watersheds of Davao and Cotabato, including the cluster of Sarangani islands. Originally, the ***B'Laan*** were not highland dwellers. Their presence in the mountain is a result of later migration.

Figure 46 The Gumne House of the B'Laan Tribe

The traditional home, *Gumne*, is the multilevel house of the ***B'Laan*** family from Datal Tampal, Malungon, in Sarangani. A typical *Gumne* comprises six to seven platforms built on four to five levels. The first platform is the *ba del*, which serves as a vestibule where rice is pounded and leads to the next platform called the *ganas* - the dining and living areas. The *ganas* lead to two other platforms: the lower platform, the same level as the *ba del*, is the *abo* where food is prepared and cooked, while the upper platform, called the *Iwang*, is where important guests and visitors are received. Adjacent to the *Iwang* are two higher platforms called the *snefil* or bedroom, where important belongings are kept. Iwang and snefil have wide hopper windows called tanbih sulong, which also serve as balconies during the day. The last platform is called the *fantas*, an attic where young women of the house sleep and are kept before marriage.

The *Gumne* traditionally faces towards the east, as the rising sun is believed to bring positive energy and good fortune to the household. Sugarcane stalks are tied on every front post for happiness and peace, while stones are placed in front of the house to signify solidarity.

The materials used in building the *Gumne* are relative to a family's social status. Those from the royal family use *sinasa* or flattened bamboo for their walls and roof, while those from the lower classes use wild sugarcane leaves. The latter usually transfer from one place to another for food and better living spaces. Because they often move out, the lower-class ***B'Laans*** do not build permanent dwellings.

Mandaya

The ***Mandaya*** tribe dwells along with the mountain ranges of Davao Oriental. They occupy upstream areas in highly dispersed settlements where they practice slash-and-burn cultivation.

Their traditional house, Bal'Lay, is rectangular in a structure erected on stilts five feet above the ground and usually occupied by two or three families. The walls are usually made of a *sayapo* bar, securely tied in bamboo laths of *uway* or rattan strips in a zigzag pattern. The tips of the laths are carved with a distinct figure called the *ligpit* to prevent joints from slipping. Posts and beams are all timber, while the walls are made of *tambullang* (flattened bamboo slides), *sinansan* (woven rattan slats), sawali (flattened tree barks), or *inak-ak* (wooden strips). The stairs, called *nuknukan,* are made of round timber, carved with distinctive foothold notches, while the handrails, called *kal'lubabay*, are installed afterwards. A *Bal'Lay* with a distinctive split bamboo roof is commonly called a *lyupakan*.

Figure 47 The Bal'Lay House of the Mandaya Tribe

Moro Traditional Architecture

Figure 48 Pre-colonial Moro Mosque at Lake Lanao, Mindanao

Finally, we look at the Muslims in Mindanao, whom the Spaniards initially dubbed the Moro people. Here, we see a Pre-colonial Moro Mosque in the region of Lanao Lake in Mindanao. The Moro people or Bangsamoro people refer to the 13 Muslim-majority ethnolinguistic Austronesian groups of Mindanao, Sulu, and Palawan, native to the region known as the Bangsamoro (lit. Moro nation or Moro country). As Muslim-majority ethnic groups, they form the largest non-Christian population in the Philippines.

Chapter 3
Miscellaneous Philippines Folk Architecture

The Maharlika History

The Land of Maharlika was composed of the Philippines, Borneo, Guam, Marianas Islands and Hawaii and was ruled by a certain King, Luisong Tagean [later changed to Tallano for fear of Spanish execution]. The Maharlika (freeman or freedman) were the feudal warrior class in ancient Tagalog society in Luzon, the Philippines. The Spanish translated the name as Hidalgos (or libres). They belonged to the lower nobility class, similar to the Timawa of the Visayan people. In modern Filipino, however, the word referred to aristocrats or royal nobility, which was restricted to the hereditary Maginoo class.

The term maharlika is a loanword from Sanskrit maharddhika (महर्द्धिक), a title meaning "man of wealth, knowledge, or ability". Contrary to modern definitions, it did not refer to the ruling class but to a warrior class (minor nobility) of the Tagalog people, directly equivalent to Visayan timawa. Like timawa, the term also has connotations of "freeman" or "freed slave" in Filipino and Malay languages. (Ref: https://en.wikipedia.org/wiki/Maharlika#Etymology)

Figure 49 Bahay Kubo in the Maharlika series

The Maharlika Inspiration

The following illustrations are inspired by paintings by JBulaong 2020.

Figure 55 Inspired by a Pre-colonial Maharlika Lakeside House painting by JBulaong 2020 at Lake Sebu, Mindanao

Figure 52 Inspired by a Pre-colonial Maharlika Mountain House painting by JBulaong 2020

Figure 51 Inspired from a Maharlika Datu House Painting by JBulaong 2019

Figure 53 Inspired by a JBulaong 2020 painting of a Pre-colonial Maharlika Split Level Upland House

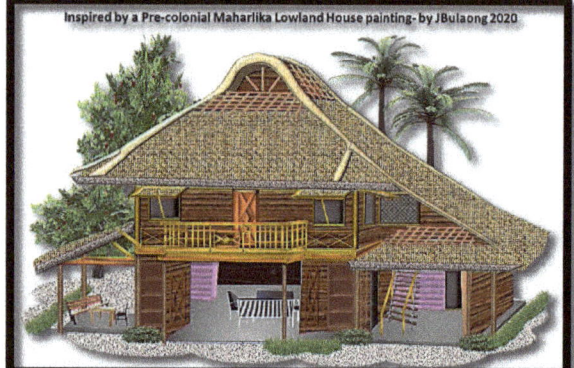

Figure 54 Inspired by a Pre-colonial Maharlika Lowland House painting- by JBulaong 2020

Figure 50 Maharlika Bee House on Hill Slope

In some Indo-Malayan languages, as well as the languages of the Muslim areas of the Philippines, 0he cognates *mardika*, *merdeka*, *merdeheka*, and *maradika* mean "freedom" or "freemen" (as opposed to servitude). The Malay term *mandulika* also meant "governor". The name of the Mardijker people of Batavia also came from the same etymon and referred to formerly enslaved people and servants under Dutch rule who were composed largely of Portuguese-speaking Catholic Goans, Moluccan Merdicas, and Filipinos (the Papangers) captured by Moro raiders.

In modern Filipino, however, the word referred to aristocrats or royal nobility, which was restricted to the hereditary Maginoo class. The next illustration below is a Datu[14] House in Mindanao.

Maharlika Grain

Maharlika. A Filipino favourite - the Maharlika rice is known for its fluffy "maalsa" feature. Its dry, hard, long-grained quality makes it perfect for cooking fried rice. It is only obvious that a Maharlika Grain Store would exist, and this illustration depicts a pre-colonial version.

Figure 56 Maharlika Grain store Adopted Bee House

Bee Shacks in vernacular fashion on Marinduque Island

Figure 57 Bee Shacks done by Agriculture Coops in Marinduque Island

FEMACO, or Federation of Marinduque Agriculture Cooperatives, is a provincial-level federation of primary agriculture cooperatives. They have 26 primary cooperative members from Buenavista, Torrijos and Mogpog municipalities. Stingless bees are a major enterprise of its primary cooperative members.

---

[14] Datu is a title which denotes the rulers (variously described in historical accounts as chiefs, sovereign princes, and monarchs) of numerous indigenous peoples throughout the Philippine archipelago.

Vernacular architecture in The Zamboanga peninsula

Figure 58 A small Subanon village hut on Mount Malindang

The Subanon[15] (also spelt Subanen or Subanun) is an indigenous group to the Zamboanga peninsula, particularly in Zamboanga del Sur and Misamis mountainous areas Occidental, Mindanao Island, Philippines. The name is derived from the word soba or suba, a word common in Sulu, Visayas, and Mindanao, which means "river", and the suffix -nun or -non, which indicates a locality or place of origin. Accordingly, the name Subanon means "a person or people of the river". The Subanon people speak Subanon languages. These people originally lived in low-lying areas. However, disturbances and competition from other settlers like the Moros and migrations of Cebuano speakers to the coastal areas attracted by the inviting land tenure laws further pushed the Subanon into the interior.

Subanons generally refer to themselves as the gbansa Subanon, meaning "the Subanon nation". They distinguish themselves from each other by their roots or point of origin. These are based on the names of rivers, lakes, mountains, or locations. The groups that traditionally remained animists call themselves Subanen in the area closer to Zamboanga City. Outsiders often call the Subanon Subano, a Spanish version of the native name.

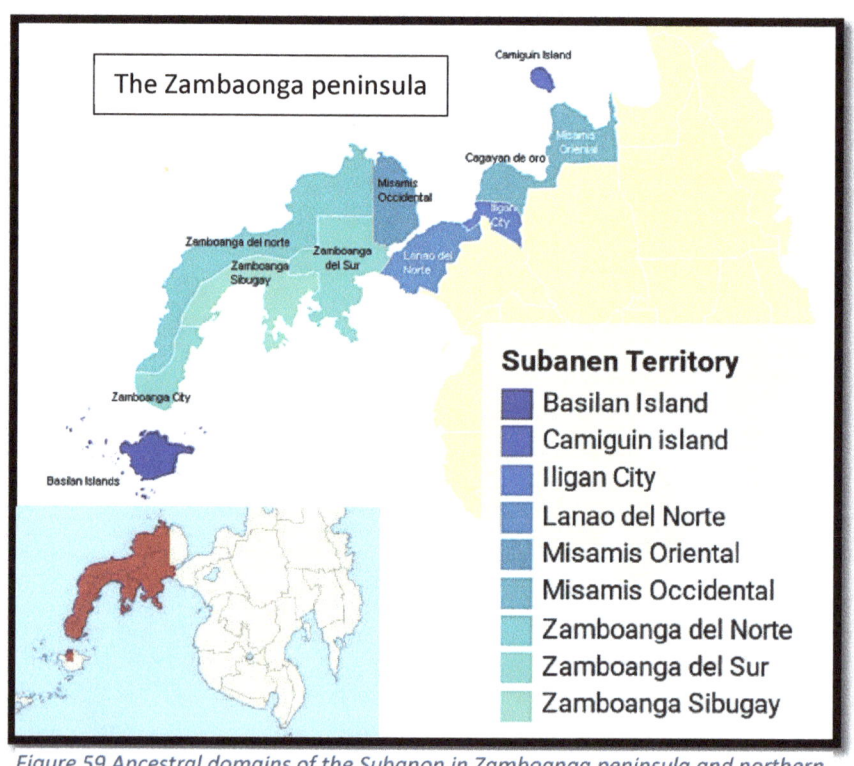

Figure 59 Ancestral domains of the Subanon in Zamboanga peninsula and northern Mindanao

---

[15] https://en.wikipedia.org/wiki/Subanon_people

Early history

For some time before the Spaniards came during the period of colonial rule, the Subanon had trade contacts with the Tausug, Maguindanao and the Maranao. As they were under the protection of the Sultanate of Maguindanao, they also provided materials, warriors and help in the war efforts of the Sultanate. They were also entitled to share in the war spoils.

The Zamboanga peninsula (https://en.wikipedia.org/wiki/Zamboanga_Peninsula)

Zamboanga Peninsula (Tagalog: Tangway ng Zamboanga; Chavacano: Peninsula de Zamboanga; Cebuano: Lawis sa Zamboanga) is an administrative region in the Philippines, designated as Region IX. It comprises three provinces (Zamboanga del Norte, Zamboanga Sibugay and Zamboanga del Sur), including four cities (**Dapitan**, Dipolog, Pagadian, Isabela) and the highly urbanized Zamboanga City.

Bee species related to the Zamboanga Peninsula:
Tetragonula dapitanensis (Cockerell, 1925)

Trigona dapitanensis Cockerell 1925a: 492: Holotype (BMNH 17b.1135). Here considered a valid and larger form[16] of laeviceps from the Philippines (taxonomy); **Type locality**: PHILIPPINES "**Dapitan**, Mindanao (Baker, 23119)" (worker);

Figure 60 Political map of the Zamboanga Peninsula, Philippines. Showing Zamboanga del Norte, Zamboanga del Sur, Zamboanga Sibugay, Isabela City, and Zamboanga City.

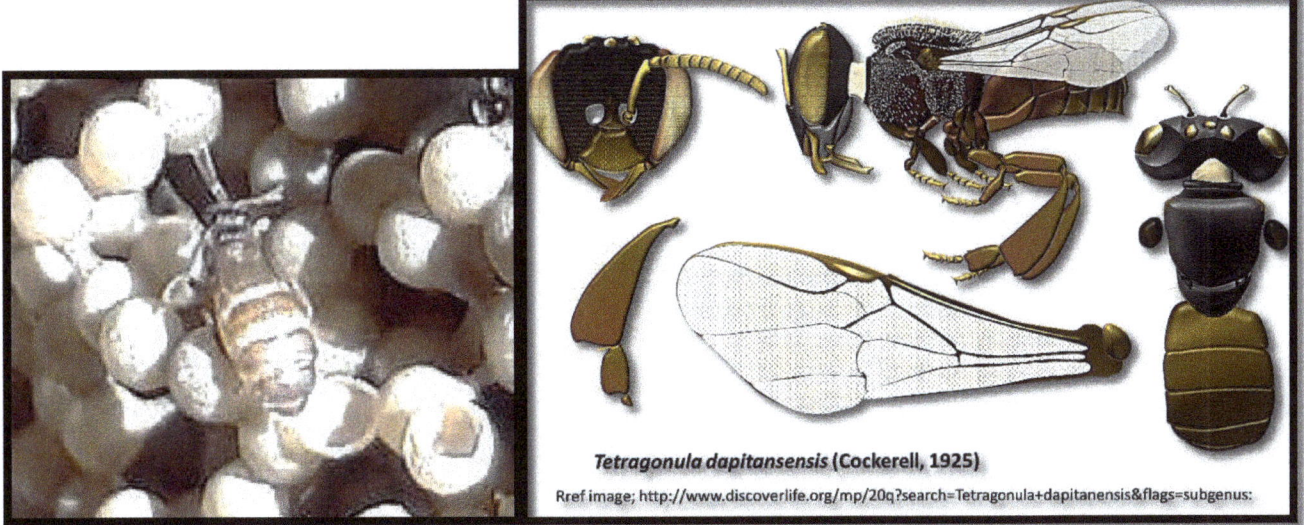

Figure 61 Tetragonula dapitanensis queen and worker illustration

---

[16] Editor's note: relative   a form is not a valid term, since it is not a subspecies or variety (which, increasingly, are not considered valid taxonomic entities)..

Vernacular architecture of Palawan

The Tagbanwa[17] people[18] are one of the oldest ethnic groups in the Philippines, mainly in central and northern Palawan. They are a brown-skinned, slim, and straight-haired ethnic group. Research has shown that the Tagbanwa are possible descendants of the Tabon Man, thus making them one of the original inhabitants of the Philippines.

Figure 62 A typical Tagbanwa hut

Bee species native to **Palawan**: *Lepidotrigona palavanica* (Cockerell, 1915)

Trigona palavanica Cockerell 1915: 2-3: Holotype (BMNH 17b.1120). The nitidiventris species group (comparative notes, taxonomy); **Type locality**: PHILIPPINES "**P. Princesa, Palawan** (Baker coll 3839)"(worker);

Figure 63 Palawan Stingless Bees cited in Cockerell 1919

---

[17] https://en.wikipedia.org/wiki/Tagbanwa
[18] https://en.wikipedia.org/wiki/Peoples_of_Palawan

Chapter 4

Vernacular Bamboo Houses in The Philippines

1. Lucia Gadian of Brgy. Awis, Leon

The house was a typical Tausug traditional Filipino bamboo architecture. There were stingless bee nests in almost every node of the bamboo wall panel. 109 in total

Figure 64 The ancestral home of Lucia Gadian of Brgy. Awis, Leon, Panay Island.

2. Antonio Capirayan of Brgy Tacuyong Sur, Leon, Iloilo

60 yr. Old Traditional Bamboo with the Tausug Influence. Bees love the Bamboo wall panel frames' nodes (hollow stems).

Figure 65 The 60 yr. old bamboo house of Antonio Capirayan of Brgy Tacuyong Sur, Leon, Iloilo, Panay – almost collapsing.

3. Esperidion Calaor ng Brgy. Gumboc, Leon, Iloilo

Another Traditional Bamboo house has 24 colonies

Figure 66 Bamboo house of Esperidion Calaor ng Brgy. Gumboc, Leon, Iloilo, Panay also houses 24 bee colonies.

A bee hut in Romblon

Joe Asis of Romblon contributed his version of Bahay Kubo, an elevated floor made of small logs and bamboo. The roof is made of coconut leaves. to maximize the space under the elevated floor, used as a shelter for domesticated animals like chickens and a four-legged security guard (his male dog named Guapito)

Balay it kiwot (bee hut) is located on Joe's farm beside rice fields and was built by *Taga Uring* (charcoal men). That's why the railings are made of slices of mangrove wood. His village is Guinbirayan in Santa Fe, Tablas Is., Romblon Province. Another term for bee hut is "*bayay Ng kabuyay.*"… "*Bayay ng Kabuyay*" is Aklan's dialect and means *Bahay ng Buyog* (bee hut).

Figure 67 A bee hut in Romblon

A rest hut at TESDA RTC garden, Iloilo

Figure 68 A rest Hut at TESDA RTC garden, Iloilo

Ms. Nora Alanza Peña contributed to Iloilo, Panay Is., where she runs a bee sanctuary and conducts training for bee product development. She had a big Nipa hut (similar to the TESDA rest hut above) collapse last week because of typhoon Agaton. It collapsed flat on the ground, and that small beehouse also fell. Her son Coby re-erected it. For her bee hut she calls it "*tradisyunal na pabahay sa bubuyog/ kiwot*"

Figure 69 Nora Peña's Bee Sanctuary and a bee rack in Leon, Iloilo

Sagada House in Tublay, Benguet, Central Luzon.

In this instance, the tour guide was contributed by Bernard Anciado and Leo Kimbungan in traditional garb. It's amazing how they preserved this 400-year-old last traditional house in Sagada. Another

Figure 70 Sagada House in Tublay, Benguet, Central Luzon.

amazing thing about it is that it looks small, but it's a three-level house and big inside.

Bee Housing by Cebu Stingless Bees

Figure 71 Bee housing by Cebu Stingless Bees.

Cebu Stingless Bees. - A Facebook community group All about bees and their food, like pollens and sources of nectars, plants, and trees. Suppose you have problems with bees in your area. Let me help you transfer or relocate their colony (*ligwan*, *kiyot*, giant bee). The three cylindrical hives are by Jifferson Camaling (Who studied at the University of Cebu Lapu-Lapu and Mandaue)

Gawahon Eco Park – in Victorias, Occidental Negros

The Eco Park Custodian, Freddy Lozada, and a few beekeepers have Stingless bee farms in their residence vicinity. Below is a Bahay Kubo on the left, an ornamental bamboo hut at Nisol Farm, Gawahon, and an Okinawa house with a Japanese accent in the Eco Park. Though the vernacular structures are not yet used as bee housing, they represent the unique design of the vernacular concept.

Balai Buhay sa Uma in Bulusan, Sorsogon.

Figure 72 Images are a Bahay Kubo on the left, an ornamental bamboo hut at Nisol Farm, Gawahon and an Okinawa house giving a Japanese accent in the Eco Park.

One of the more impressive bee farms in the Philippines. They have more than 3,000 stingless bee colonies. Many of their structures are vernacular conceptually.

Novelty bee housing in Los Pepes, Cavite

Los Pepes is run by Christ Mark Bergonio in Indang, Cavite

Figure 73 Rest hut in Balai Buhay sa Uma

Figure 74 Novelty bee housing in Los Pepes, Cavite; Marl's Bee store housing his feral colonies

Ilog Maria Honeybee Farm - Silang, Cavite.

Here, I observed conical roofed round houses, though two of them are restrooms, and the big one in the middle is a Bee gallery. Some indigenous people may have influenced such round or octagonal houses in Luzon.

Figure 75 Left and right images are restrooms and the big one in the middle is a Bee gallery.

Norma B & B San Cristóbal in Oriental Mindoro has a new spread for visitors.

Feral colonies: Such is the situation in which a stingless bee's nest occupies almost any corner or opportune cavity. They enjoy reproducing and making nests in abandoned cardboard boxes and the foundation cracks and walls in every conceivable spot in the bamboo structure and joints. Even in pipe elbows and drainage fixtures, as seen in the photo below.

Figure 76 Rest huts made of bamboo and we found several nest entrances built in the joints and the bamboo structures.

Philippines Honeybees Industries (PHI) Bibera Research Training Centre in Pilar, Sorsogon, Bicol Region, Managed by Mr. Domingo. We stopped by to observe progress on Kiyot hive propagation by Coconut shell method from Fern hives obtained from the wild. Coconut shell hives are propagated under the cool protection of fan palm attap roofing, neatly arranged by traditional methods.

Figure 77 Philippines Honeybees Industries (PHI) Bibera Research Training Centre in Pilar, Sorsogon, Bicol Region,

Grajos Beefarm in Sorsogon.

Leo Grajo builds a hexagon beehouse that allows bees to fly out for foraging in six directions. This provides for a better colony success rate. The top portion is elevated to provide ample ventilation.

Figure 78 Hive racks and a Hexagon Bee House.

Pia's Bee Farm, Lipa City, Batangas by Lee Gaitana

I was in Pia's Bee farm in 2018 on a Sothern Luzon tour. It is owned by Lee Gaitana and was named after his only daughter, Pia, who is growing up exposed to beekeeping. She spent most of her time on the farm or with her father during beekeeping consultancy visits from infancy to toddler years. Pia is the namesake of Padre Pio, her patron saint. The well-manicured garden is pleasantly reminiscent of my Beescape book. What caught my attention was the rustic structure that housed his Bee Gallery. It is an eatery cum gallery on two levels.

Figure 79 The Bee Gallery at Pia's Bee Farm and the two-level structure by night.

Honey House Bees & Beeks in Apokon Road in Tagum City.

Figure 80 Owners are Peter Paul Rafasola and his wife Claire Azuelo-Rafasola are former beekeepers in America. The box hive construction appears to have an Australian design influence with a clear plastic cover for observation. Encountered Tetragonula hive in a woven bamboo wall.

We encountered a *Tetragonula* hive in the woven bamboo wall. The box hive construction appears to have an Australian design influence with a clear plastic cover for observation. This farm also breeds Apis melifera together with Tetragonula biroi.

The owners are Peter Paul Rafosala and his wife, Claire Azuelo-Rafosala, former beekeepers in America. They started in Tagum City in 2014 and specialize in bee pollen, comb honey, propolis and bees wax. They also offer hive equipment.

Mikee Balais in Tacloba City, Leyte Island

Figure 81 Mikee was badly hit by the Typhoon, as shown by his Aunty Feli

Mikee lives with his kind aunt, Fely Sommer… located somewhere behind the Robinson Marasbaras in Tacloban City. Mikee took us to the back of the house where his previous Meliponay was devastated after a typhoon at the end of 2019. Then he was at his auntie's place, who showed us some photos of the actual destruction.

One of his colonies was bred in a PVC pipe...To our great surprise, the beehive was opened for us to take a closer look when we saw the alarming situation in the colony due to the creeping moulds at the brood. This mould usually happens due to constant heavy rains, and the PVC material does not allow normal dissipation of accumulated humidity.

Figure 82 Mikee Balais' Meliponary in traditional architecture with a 'green roof' for his shack.

Fast forward to the current situation; he has since upgraded his place. Mikee Balais' Meliponary has an A-frame chalet in traditional architecture with a 'green roof' for his adjacent shack. The green roof appears to be some large leaf cucurbit-type creeper. A marvel idea to provide favoured nectar for his bees while keeping the shack cool

The Kalinga Octagon Bee House

An example of the octagon house is the Kalinga tribe in Luzon.

Figure 84 Example of an octagon house of the Kalinga tribe in Luzon

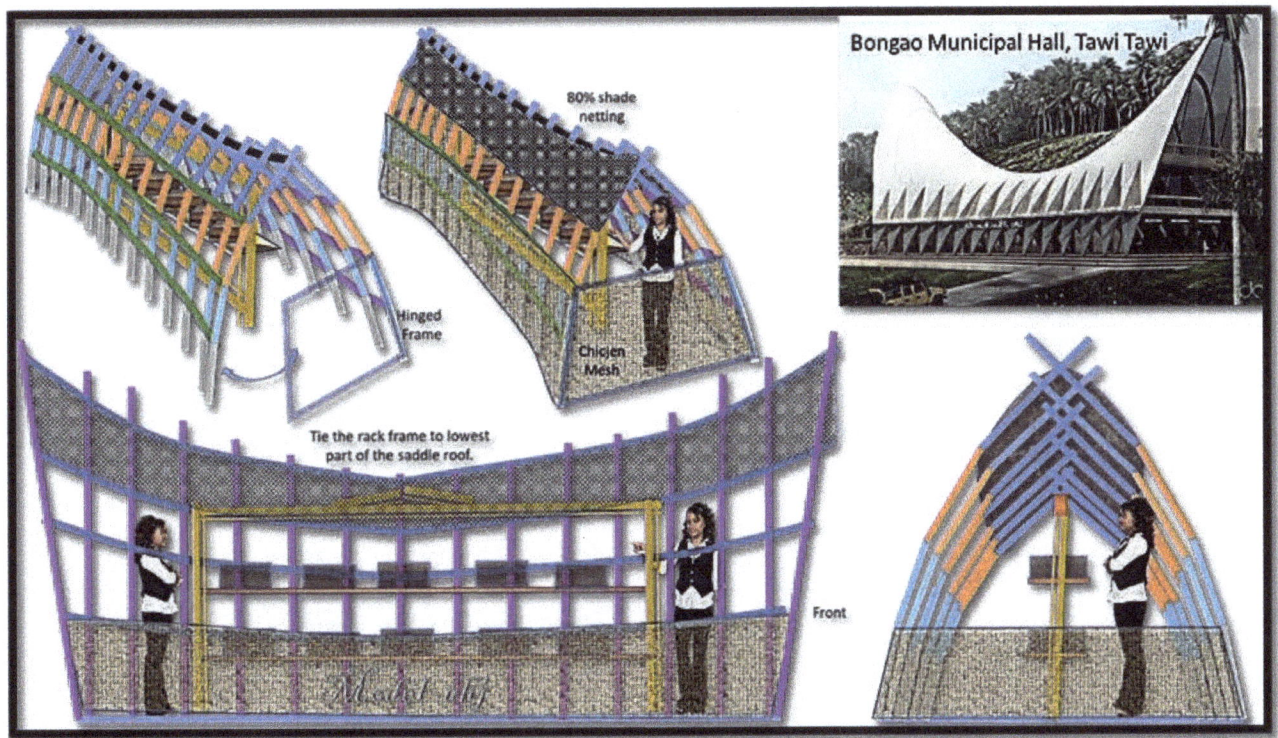

Figure 83 Inspired by the Bongao Municipal Hall, Tawi-Tawi

Addendum - Vernacular Filipino Construction Terms.

A

Adobe Anchor -*Liyabe*

Alignment -*Asintada*

Aligned -*Kalinya*

Alternate Or Staggered -*Unosinotra*

Anchor -*Liyabe Astragal -Batidura*

Awning -*Media Agua*

B

Balusters -*Baluster, Barandilla*

Banisters -*Barandillas*

Barrel Bolt -*Trankilya*

Baseboard -*Rodapis*

Bath Tub -*Baniera*

Brace -*Pie De Gallo*

Brick -*La Drillo*

Beam -*Biga*

Bolt -*Pierno*

Bottom Chord -*Barakilan*

Bottom Chord -*Tirante*

C

Canopy -*Media Agua*

Cabinet Hinge -*Espolon*

Carpentry -*Karpinteria*

Cast Iron -*Pundido*

Ceiling -*Kisame*

Ceiling Joist -*Kostilyahe*

Cement Brick -*Ladrilyo*

Cement Tiles -*Baldosa*

Chain Bolt -*Baral De Kadena*

CHB (**Concrete Hollow Block**) Laying -*Asinta*

Closed Stringer -*Madre De Eskalera*

Collar Plate -*Sinturon*

Column -*Kolumna, Haligue*

Concrete Slab -*Larga Masa*

Conductor -*Alulod*

Contactor -*Kontratista*

Corner -*Kanto*

Corrugated G.I. Sheets -*Yerocanalado Galbanisado*

Crushed Stone -*Eskondro*

D

Diagonal Brace -*Pie De Gallo*

Diagonal Brace -*Sinturon*

Diglead -*Tingga, Estopa*

Dish Rack -*Bangguerahan*

Door Fillet -*Batidura*

Door Head -*Sombrero Pintuan*

Door Jamb -*Hamba Pintuan*

Dowel -*Abang*

Down Spout -*Tubo De Banado*

Drawbore Pinor -*Punsol*

Draw Pin -*Punsol*

E

Earthfill -*Eskumbro*

Eave -*Sibe*

Eave -*Alero*

External Siding -*Tabike*

Electrician -*Electrisista*

F

Fascia Board -*Senepa*

Faucet -*Gripo*

Feet -*Piye*

Fill -*Tambak*

Filler -*Tapal, Dagdag*

Fillet -*Batidura*

Flooring -*Sahig, Suelo*

Floorboards -*Dotal*

Floor Joists -*Soleras*

Floor Sill -*Guililan*

Flush -*Alahado*

Foreman -*Kapatas*

Frame Work -*Balangkasg*

G

Gable Roof -*Dos Aguas*

G. I. Strap -*Lingueta*

Girder -*Kuling*

Girt -*Sepo*

Groove -*Kanal*

Gutter -*Kanal*

H

Hall -*Caida*

Handrail –*Gabay*

Hinge –*Bisagra*

Horizontal Stud –*Pabalagbag*

I

Inches –*Pulgada*

Iron -*Hiero*

J

Jamb -*Hamba*

Joist –*Suleras*

L

Labourer –*Piyon*

Lean-to Roof –*Sibe*

Low Table –*Dulang*

M

Mason -*Kantero*

Masonry Fill –*Lastilyas*

Miter -*Canto Mesa*

Mix Of Cement &Gravel -*Lastilyas*

Mortar -*Paupo*

Mortar Joints -*Kostura*

Moulding -*Muldura*

N

Nail Setter -*Punsol*

Nailers -*Pamakuan*

Nails -*Pako*

Newel Post -*Tukod*

Nicolite Bar -*Estanyo*

Nut -*Tuerka*

O

Oakum -*Estopa*

Open Stringer –*Hardinera*

Overhang Or Projector -*Bolada*

P

Panel Door –*De Bandeha*

Pattern Or Schedule -*Plantilya*

Pea Gravel -*Grabita*

Pendulum (King Post) -*Pendulon*

Pick Work On Masonry –*Piketa*

Plain G.I. Sheet -*Yero Lisogalbanisado*

Plain G.I. Strap -*Lingueta*

Plank Board -*Senepa*

Plaster -*Palitada*

Plastered Course –*Kusturada*

Plug -*Tapon*

Plumbing -*Tuberia*

Plumb Line -*Hulog*

Plum Bob -*Hulog*

Post -*Halige / Poste*
Projection –*Bolada*

Purlins -*Reostra*

Putty -*Masilya*

Q

Quarter Round -*Media Cana*

R

Rabbet -*Vaciada Rafters -Kilo*

Reinforcing Bar -*Cabilla Bakal*

Ridge Roll -*Caballete*

Riser -*Senepa, Takip , Silipan Rivets -Rimatse*

Roof -*Atip, Bubong*

Riser –*Takip Silipans*

S

Scaffolding -*Andamiyo*

Scratch Coat -*Rebocada*

Screw -*Turnilyo*

Sheet -*Plantsa*

Shower -*Dutcha*

Siding (External) -*Tabike*

Sink -*Prigadero*

Sketch Plan -*Krokis*

Slab (Rough) -*Larga Masa*

Slope -*Bahada*

Solder -*Hinang*

Solder Bar -*Estaniyo*

Soldering Lead -*Estanyo*

Spacing Of Gap -*Biento*

Stake -*Estaka*

Stringer (Closed) -*Madre (De Escalera)*

Stringer (Open) -*Hardinera*

Stucco Or Plaster -*Palitada*

Stud (Horizontal) -*Pabalagbag*

Stud (Vertical) -*Pilarete*

T

Temper(metal Work) – *Suban, Subuhal*

Tinsmith -*Latero*

Trellis -*Pergola*

Truss -*Kilo*

Top Chord -*Tahilan*

Tread -*Baytang*

V

Varnish Finish -*Monyeka*

Vertical Stud –*Pilarete*

W

Wainscoting Tiles -*Asolehos*

Wall Post –*Bagad*

Washer -*Pitsa*

W. I. Strap -*Planchuela*

Window Head -*Sumbrero*

Window Sill -*Pasamano*

Wiring Knob –*Poleya*

Window Or Door Jamb – *Hamba*

Wood Grain –*Haspe*

Wood Plank –*Tabla*

Wrought Iron Strap – *Plantsuwela*

VOLUME 3, PART 2
~ OCEANIA OUTLOOK~

Prologue to Part 2

Here, we can pass through time to glimpse a period near the end of European colonialism. Oceania was explored by Europeans in the 16th century onward. Portuguese explorers, between 1512 and 1526, reached the Tanimbar Islands, some of the **Caroline Islands** and west Papua New Guinea. On his first voyage in the 18th century, James Cook came through and later arrived in Australia for the first time.

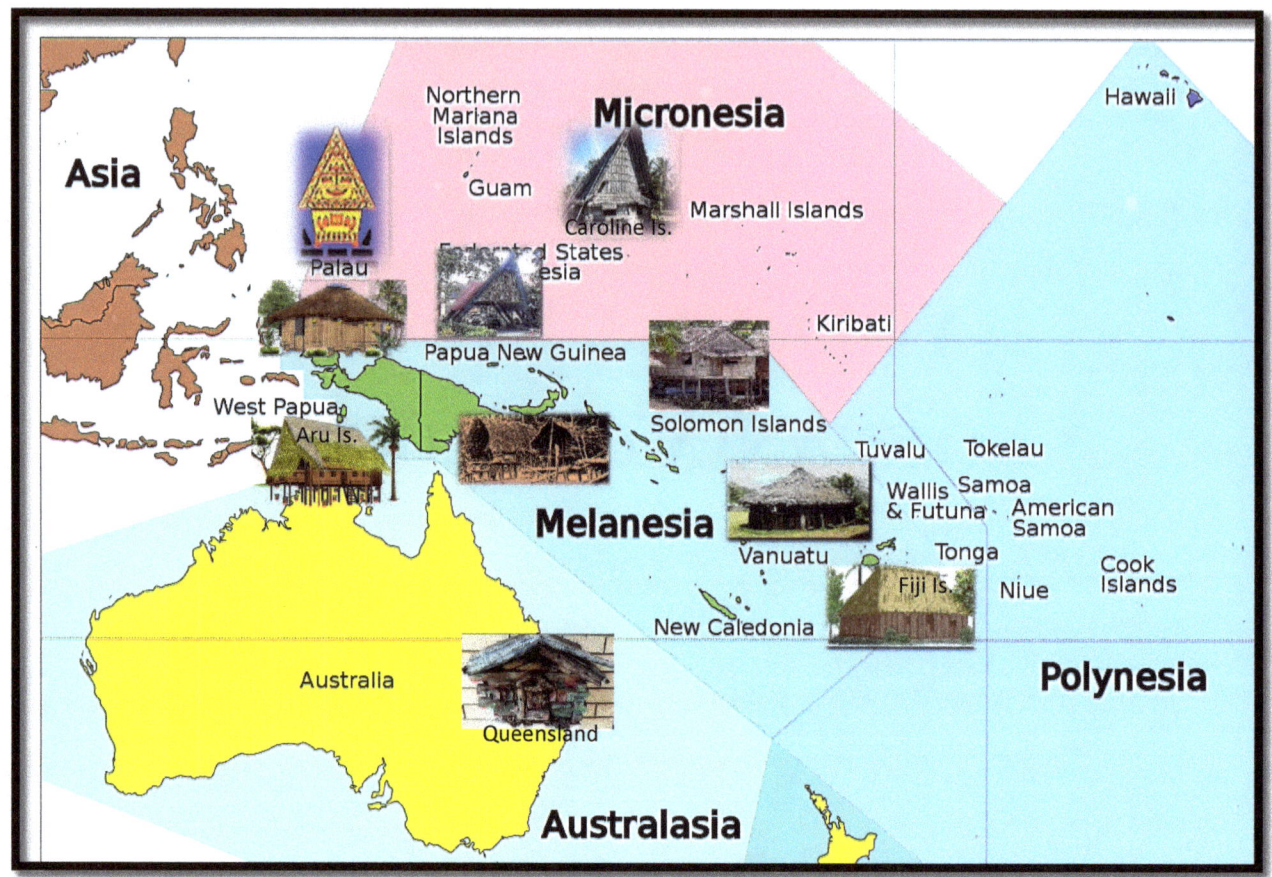

Figure 85 Subregions of Oceania.

The arrival of European settlers in subsequent centuries (driven by the lucrative spice trade of The East Indies) significantly altered Oceania's social and political landscape. This part will cover regions depicted in historical records of localities where stingless bees were discovered. The Oceania regions mentioned are: 1). Australasia – Australia; 2) Melanesia – West Papua (Irian Jaya), Papua New Guinea, Solomon Islands, Vanuatu & Fiji; and 3). Micronesia – Palau and Caroline Islands. The country of Timor Leste (included in this part), although geographically in the Indonesian Archipelago, is politically autonomous and independent of Indonesia.

On the other hand, the Aru Islands and West Papua are in The Republic of Indonesia but geographically on the Sahul shelf and in the region of Melanesia. This position likely permits the existence of the insular SB genus *Sahulotrigona* on these Islands.

Native houses[19] in Dobbo, **Aru Islands, East Maluku**

Figure 86 Native houses in Aru Islands

Map 1 Aru Islands

The Aru Islands Regency (Indonesian: Kabupaten Kepulauan Aru) is a group of about 95 low-lying islands in the Maluku Islands of eastern Indonesia.

Meliponine distribution in Micronesia

Stingless bee records relevant to Micronesia:
Tetragonula fuscobalteata (Cameron, 1908) Schwarz 1939b: 151-152 (distribution, taxonomy); **PALAU ISLANDS**, Arukoron; Babelthaob, Marukyoku-Kaisharu; **Korror**; Schwarz 1948: 40, 56, 69, 83, 90, 118; Krombein 1950: 102, 133, 135 (distribution, morphology); MICRONESIA, **Caroline Islands.**

Figure 87 Left: Hoise emblem of Koror State Flag; Right: Island of Yap, Federated States of Micronesia, Caroline Islands, Pacific

Tetragonula valdezi (Cockerell, 1918)
Schwarz 1939b: 151, 152 (distribution, variation); MICRONESIA, **Caroline Islands**, Toloas; Truk; Cockerell 1939a: 61, 62, 64 (distribution ("presumably introduced by man"), variation); MICRONESIA, Dublon Island; Tarik Island; Schwarz 1948: 118, 233; Krombein 1950: 102, 133, 135 (distribution, morphology); MICRONESIA, **Caroline Island**s;

Republic of Palau

Palau[20], officially the Republic of Palau and historically Belau, Palaos or Pelew, is an island country and microstate in the western Pacific.

Etymology

---

[19] Inspired by a photograph taken during the voyage of H.M.S. Challenger (1872-1876), funded by the British Government for scientific purposes.
[20] https://en.wikipedia.org/wiki/Palau

The name for the islands in the Palauan language, Belau, derives from the Palauan word for "village", beluu (thus ultimately from Proto-Austronesian *banua), or from aibebelau ("indirect replies"), relating to a creation myth. The name "Palau" originated in the Spanish Los Palaos, eventually entering English via the German Palau. An archaic name for the islands in English was the "Pelew Islands". **Palau**, not to be confused with *Pulau*, is a Malay word meaning "island" (found in some regional names).

Palau lies on the edge of the typhoon belt. Palau has a history of strong environmental conservation. For example, the Ngerukewid Islands and the surrounding area are protected under the Ngerukewid Islands Wildlife Preserve, established in 1956. Culturally, Palauan society follows a strictly matrilineal system (probably originated in the Philippine archipelago in the pre-colonial era).

Figure 88 Top Left: A traditional Palauan bai: Top Right: Village on the Palau Islands, painting by Rudolf Hellgrewe c. 1908.; Bottom left: Palau International Airport; Bottom Right: Flag of Koror State Palau

Tetragonula clypearis (Friese, 1909)
  Trigona laeviceps variety ***clypearis*** Friese 1909("1908"): 356, 358, fig. 15-3: Unknown; paralectotype (ZMHB (1)) (distribution, taxonomy); **Type locality**: INDONESIA (**IRIAN JAYA**) "Manikion, 14.-28. February 1903"; "Wendesi 29. Juni 1903, New Guinea" (worker); Ikudome & Kusigemati 1996: 17, 19-20 (common name (bai), distribution, nest, taxonomy); **PALAU ISLANDS**.

Chapter 5

An Australian Perspective

Moving on from arabesque and ornate façade and rake board carvings of Sundaland, we visited Australia sometime back and recorded some unique cottage designs. In contrast to filigree incorporated as beehive roof vents and pediments, we find Australian hive designs contemporary and minimalistic, and even an old telephone booth (Clark Kent's) design.

We visited Bob Luttrell in Brisbane some years back and noted well-established box hives; he has this old cottage design. We recorded this modern version of an old cottage of the early 1900s.

Figure 89 Old Australian cottage design housing Tetragonula hockingsi *in Brisbane belonging to Bob Luttrell.*

While there, I attended the Meliponine conference (Organized by Tim Heard), and they had a Bee Hive Design exhibition. The exhibition was part of the conference activities, and the local participants were encouraged to display their innovations in bee hive design.

One that caught my eye was this post-modernist (Clark Kent's) Telephone booth. I was chatting with another foreign participant; interestingly, he wanted to know whether Clark Kent was an Australian beekeeper. I just said he could be Tim Heard's 'protégé' (or was it the other way around?)

Figure 90 I recorded this modern version of a cottage. And this post-modernist (Clark Kent's) Telephone booth.

Similar design to my original "basic log cabin", I've spent a fair bit of time shaping the logs for a more natural look, and I think they've turned out okay. These split the same as a standard native bee box, so you can split into them or do Budding to get bees in. They have a viewing window, too. Available at www.hivecraft.com.au (Steve Flavel www.nativebeehives.com)

Figure 92 Log Cabin–Shaped Logs

Next is the different ethnic architecture. We shall consider whether they are purely novelties, whether the roofs' vortices assist in heat dissipation, and whether air movement over and through the unique structures will help with thermal comfort in the hives.

Figure 91 Some classical designs of Australian Bee housing.

Australian Native Stingless Bees, though not Indo-Malayan in essence, referred to as the Indo-Australasian clade:

Stingless Bees type locality relevant to Australia:

Austroplebeia australis (Friese, 1898)
Trigona australis Friese 1898: 428, 430, 431: Lectotype (ZMHB, worker): here designated, "C. Austr. / v. Müller. 93", "Trigona / australis / det. Friese 1898", "n.sp."; paralectotypes (SMNS (unknown number), ZMHB (1)) (comparative note, key to species, taxonomy); **Type locality**: AUSTRALIA "C.-Australien (v. Müller 1893)" (many workers); "Burnett-River (Queensland, Simon 1891-92)" (several workers); Cockerell 1905: 221 (comparative notes); Friese 1909: 272, 273 (citation, key to species); AUSTRALIA, Queensland, Mackay;

Figure 93 *Austroplebeia australis* (Friese, 1898)

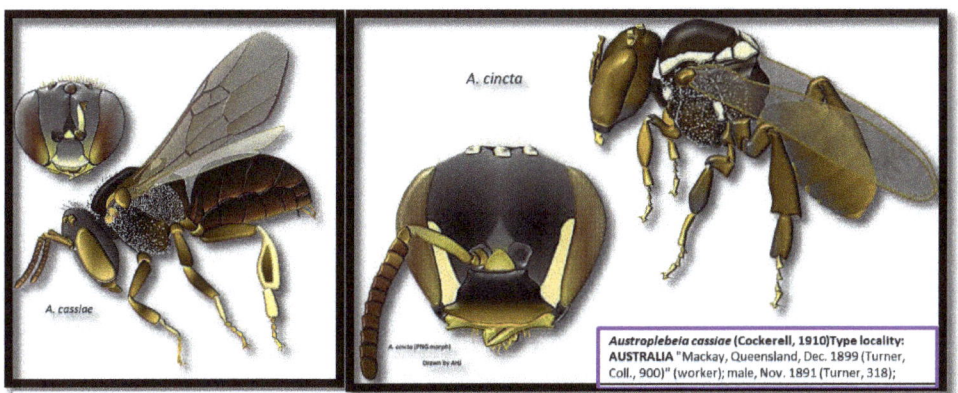

Figure 94 *Austroplebeia cassiae* (Cockerell, 1910)

Austroplebeia cassiae (Cockerell, 1910)
Trigona cassiæ Cockerell 1910: 247: Holotype (BMNH 17b.1137, worker): examined, "Type" (red border), "B.M. TYPE / HYM. / 14B.1137", "Trigona / cassiae / Ckll TYPE", "Turner Coll. / 1912-111", "Mackay / 12.99 / Cassia", "900"; paratype in "Rayment collection" (not located) (comparative notes, floral record, taxonomy); **Type locality**: AUSTRALIA "Mackay, Queensland, Dec. 1899 (Turner, 900)" (worker); male, Nov. 1891 (Turner, 318); Floral record: Cassia;

Austroplebeia essingtoni (Cockerell, 1905)
Trigona essingtoni Cockerell 1905: 220-221: Holotype (BMNH 17b.1138, worker): examined (head missing), "Syntype" (blue border), "B.M. TYPE / HYM. / 14B.1138", "Pt. Essing / ton / 42-", "Trigona / essingtoni Ckll / TYPE"; paratype (BMNH, labelled cotype (1)) (taxonomy); **Type locality**: AUSTRALIA "Port Essington, N. Australia, with no. 42.1" (2 workers);

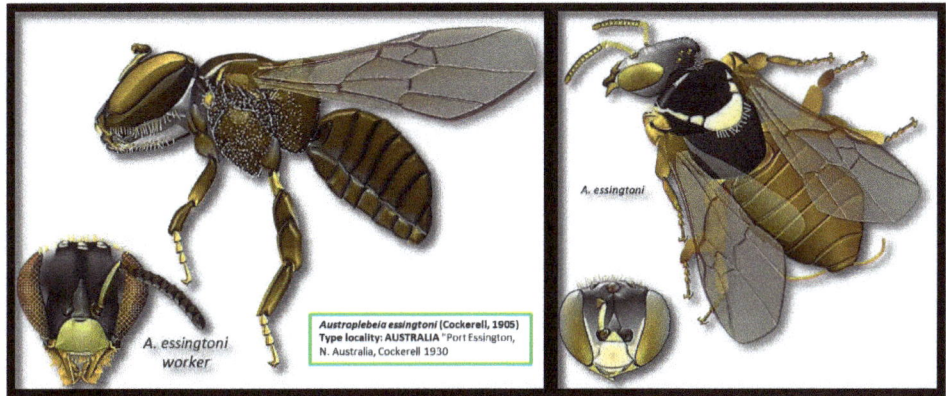

Figure 95 *Austroplebeia essingtoni* (Cockerell, 1905)

Austroplebeia ornata (Rayment, 1932)

Trigona cockerelli ornata subsp. nov. Rayment 1932f: 107: Lectotype (MVMA, worker): here designated, "4", "TYPE", "Trigona / cockerelli Raym. / ornata / C. York Q.", "SYNTYPE / T-11575 / Trigona / cockerelli / ornata", "MUS. VIC. / ENTO 2008-8 L"; Paralectotypes (MVMA (3)) (taxonomy (as manuscript name)); **Type locality**: AUSTRALIA "Cape York, Queensland (Donald Thomson)" (workers); Rayment 1935: 735 (key to species, taxonomy (described as new species again));

Austroplebeia percincta (Cockerell, 1929)
Trigona cincta subsp. *percincta* nov. Cockerell 1929e: 242: Syntypes (ZMHB, four workers): examined, labelled by Friese as Trigona cincta? (Comparative note (corrected identification of Cockerell 1910)); **Type locality**: AUSTRALIA "Hermannsberg, Finke River" (unknown);

Austroplebeia symei (Rayment, 1932)
Trigona symei Rayment 1932f: 106: Holotype (ANIC, worker): examined, "TYPE", "Trigona / symei Raym / NT. QUEENS", "AUST. NAT. / INS. COLL." (taxonomy (as manuscript name)); **Type locality**: AUSTRALIA "North Queensland" (worker);

Austroplebeia websteri (Rayment, 1932)
Trigona websteri Rayment 1932f: 105: Holotype (ANIC, worker): examined, "TYPE", "Trigona / websteri Raym. / WYNDHAM, W.A. / [below] 25/1/1931", "AUST. NAT. / INS. COLL."; paratype (ANIC (1)) (taxonomy); **Type locality**: AUSTRALIA "Wyndham, North-western Australia (U.N. Webster, M.D., 25th January 1931)" (worker);

Tetragonula carbonaria (Smith, 1854)
Syn. *Trigona angophorœ* Cockerell 1912: 225: Holotype (BMNH 17b.1139) (comparative notes, floral record, taxonomy); **Type locality**: AUSTRALIA "Sydney, New South Wales, Dec. 1, 1910 (Froggatt, 118)" (worker); Floral record: Angophora;
Trigona carbonaria Smith 1854: 414: Holotype (BMNH 17b.1136, worker) (taxonomy); **Type locality**: AUSTRALIA "Australia" (worker); Cockerell 1929e: 243 (distribution); AUSTRALIA, Ourimbah, New South Wales; Cockerell 1929d: 301 (common name (karbi), distribution, nest); AUSTRALIA, Moreton;

Tetragonula clypearis (Friese, 1909)
Syn. *Trigona wybenica* Cockerell 1929d: 300: Holotype (USNM 54960, worker): examined; paratypes (BMNH (1), QM (3), ZMHB (1)) (common name (wyben), nest, taxonomy); **Type locality**: AUSTRALIA "Thursday Island" (worker);

Tetragonula davenporti (Franck, 2004)
Previous combination - *Trigona (Heterotrigona) davenporti* Franck et al. 2004: 2325, 2326, 2327, 2328, 2330-2331, figs. 5a, 6a: Holotype (QM, worker); paratypes (ANIC (2), BMNH (2), MNHN (2), QM (2), "INRA collection, Montpellier" (8)) (genetics, illustration, microsatellite, nest, taxonomy); **Type locality**: AUSTRALIA "AUSTRALIA, QLD, Mudgeeraba, Austinville Road, V. 1990, P. Davenport leg." (worker);

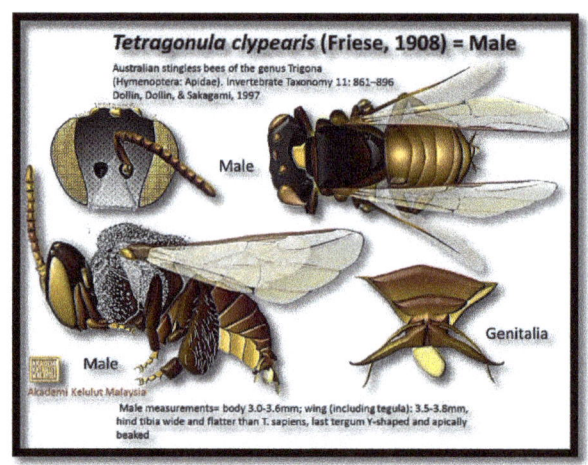

Figure 96 Tetragonula clypearis (Friese, 1909)

Tetragonula hockingsi (Cockerell, 1929)
Trigona carbonaria hockingsi n. subsp. Cockerell 1929a: 8: Holotype (QM, worker); paratypes (BMNH (2), QM (1)) (comparative note, distribution, nest); **Type locality**: AUSTRALIA "Cape York Peninsula, Harold Hockings" (five workers); "Port Darwin";

Tetragonula mellipes (Friese, 1898)
Trigona mellipes Friese 1898: 428, 429: Syntypes (ZMHB, one worker, one male): examined, "C. Austr. / v. Müller. 93", "Trigona / mellipes / det. Friese 1897", "Zool. Mus. / Berlin", "Coll. / Friese" (only worker). Friese stated the type locality was "S. Australia", probably an error for C. Australia, as ZMHB specimens match the original description and the taxon is not known from southern Australia (key to species, taxonomy); **Type locality**: AUSTRALIA "S. Australia" (worker, male);

Figure 97 Tetragonula mellipes (Friese, 1898)

T. mellipes occurs in NT and WA. *T. hockingsi* occurs in northern QLD and parts of NT, and *T. carbonaria* mainly occurs in southern QLD and NSW.

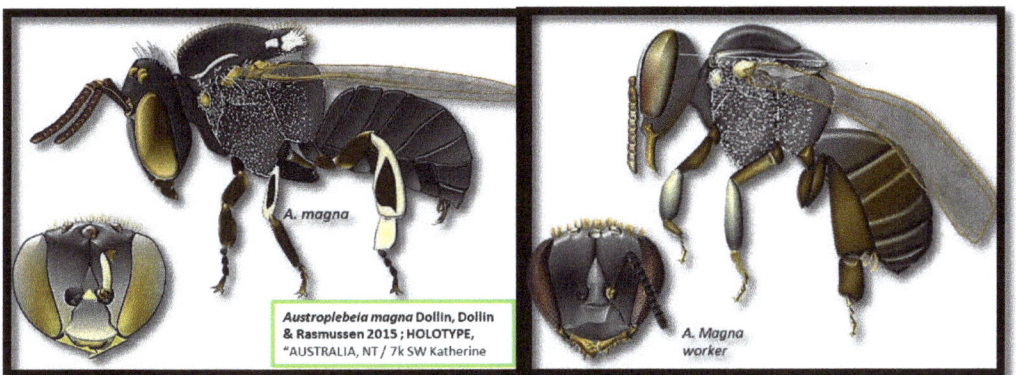
Figure 98 Austroplebeia magna Dollin, Dollin & Rasmussen 2015

Austroplebeia magna Dollin, Dollin & Rasmussen 2015 [Etym.: The Latin feminine adjective, magna, meaning 'large', refers to the broad basitarsus III and long sting lancet in workers of this species.]

Remarks. The NT populations identified as A. magna sp. nov. were part of the 'symei' group defined in our preliminary studies (Halcroft et al., 2015). Its dark colouration characterised that group. It also included the eastern species, A. cassiae. Although superficially similar to *A. cassiae* in size and colouration, A. magna workers could be distinguished by their broad basitarsus III, long sting lancet and fine clypeus hair. Source[21]

---

[21] ***Austroplebeia magna* Dollin, Dollin & Rasmussen 2015 ; HOLOTYPE, "AUSTRALIA, NT / 7k SW Katherine**

Figure 99 Distribution of Australasian and Papuasian Stingless Bees in Wallacea & Sahul

Chapter 6
Melanesia

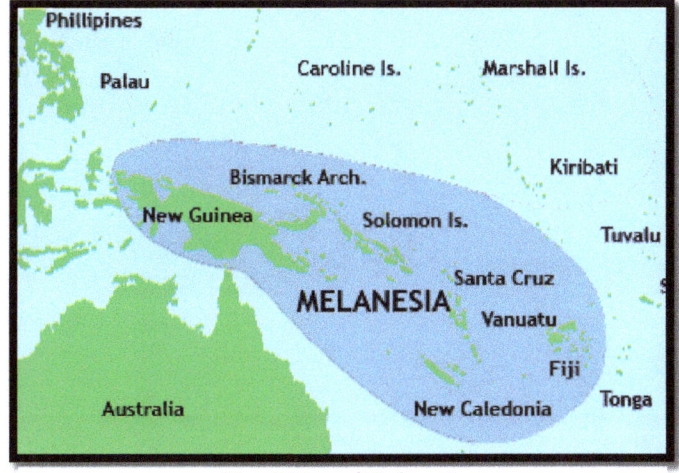

Map 2 Map of Melanesia

We look at these South Pacific Islands because an Indo-Malayan stingless bee species has its Type locality in the Solomon Islands.

The charts were prepared for a friend who intends to start a Bee Sanctuary in Misool, Raja Ampat, West Papua, where he runs a resort, and he has aspirations to house a Meliponini Breeding Center - in Waigeo/Waisai, which is also located in Raja Ampat Regency, West Papua. Using these structures depicting native heritage should give an added boost to Insect Tourism in the Area.

Bee species type locality in Melanesia: Record of bees.

Tetragonula sapiens (Cockerell 1911)
Trigona sapiens Cockerell 1911: 176: Holotype (ANIC, worker) (comparative notes, taxonomy); **Type locality**: SOLOMON ISLANDS "Solomon Islands, July-August, 1909 (Froggatt)" (worker); Cockerell 1929e: 242-243 (comparative note, distribution, taxonomy); SOLOMON ISLANDS, Lavoro Plantations, Guadalcanal Is.; New Georgia, W. Solomon; Cockerell 1936: 225 (distribution); SOLOMON ISLANDS, Halaita, Nggela; (Rasmussen 2008).

Figure 100 Tetragonula sapiens (Cockerell 1911)

Kiribati, officially the Republic of Kiribati, is an island country in Oceania in the central Pacific Ocean. The state comprises 32 atolls and one remote raised coral island, Banaba.

Figure 101 *From Left, House on stilts at the lagoon of Abaiang atoll; A house of Tebunginako village; Kiribati style church in South Abaiang; 2 storey thatched house in South Abaiang*

Solomon Islands

Figure 102 Medium shot of idyllic village beach near Auki, the capital of Malaita.

Etymology: In 1568, the Spanish navigator Álvaro de Mendaña was the first European to visit the Solomon Islands archipelago, naming it Islas Salomón ("Solomon Islands") after the wealthy biblical King Solomon. The people of the Solomon Islands were notorious for headhunting and cannibalism before the arrival of the Europeans.

Recent earthquakes in the Solomon Islands (Figure 101)

Another group of Islands in Melanesia of interest is the Fiji Islands because of its rich history and varied vernacular architecture in past lithographs (see Figure 102 *Fijian Bure*).

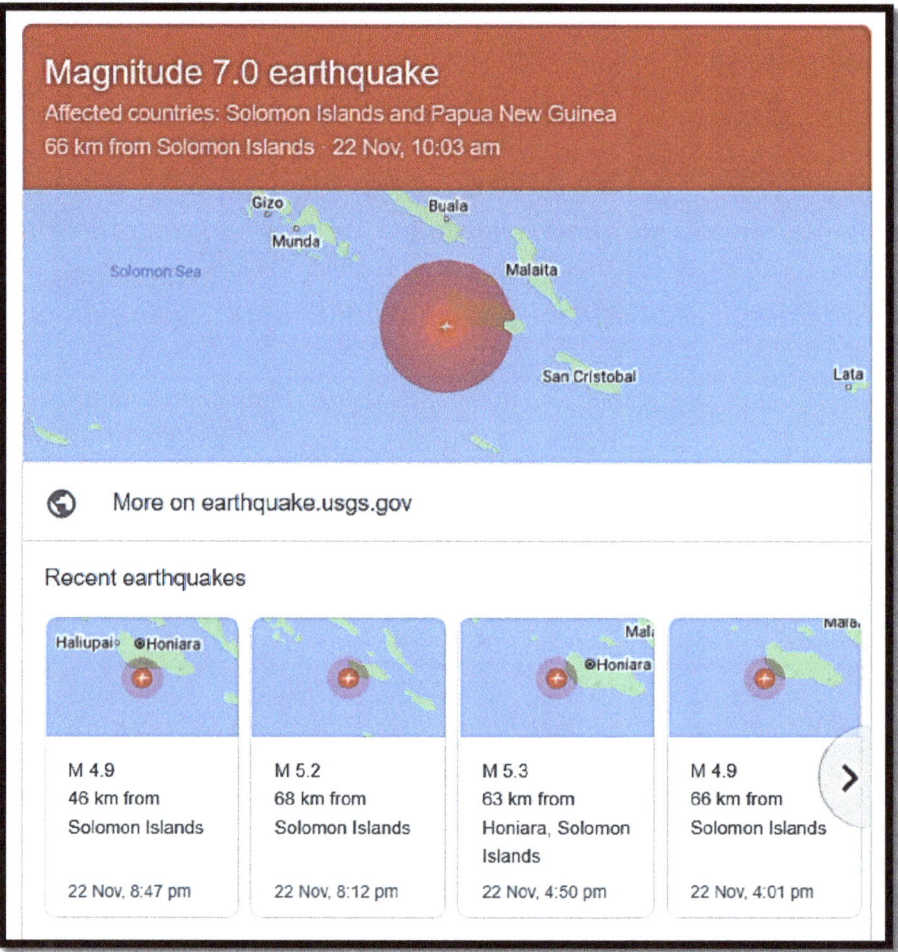

Figure 103 Recent Earthquakes on the Solomon Islands as of 22, Nov. 2022

Reviving Fiji's Traditional Architecture[22]

Figure 104 a Vale Vakaviti and a Fijian Bure

Young Fijian people of today, to those in the diaspora and those living in Fiji, need to know and appreciate their rich cultural heritage. They can read and learn their history from books and online research; however, their cultural learning will not surpass the unique experience of practically constructing and living in a Vale Vakaviti. It is critical for Fiji to revive the building of traditional houses to avoid the complete loss of traditional knowledge.

Figure 105 Levuka, 1842... The first Europeans to land and live among the Fijians were shipwrecked sailors like Charles Savage.

Etymology[23]: The name of Fiji's main island, *Viti Levu*, served as the origin of the name "Fiji", though the common English pronunciation is based on that of Fiji's island neighbours in Tonga. An official account of the emergence of the name states: *"Fijians first impressed themselves on European consciousness through the writings of the members of the expeditions of Cook who met them in Tonga. They were described as formidable warriors and ferocious cannibals, builders of the finest vessels in the Pacific, but not great sailors. They inspired awe amongst the Tongans, and all their Manufactures, especially bark cloth and clubs, were highly valued and in demand. They called their home Viti, but the Tongans called it Fisi, and it was by this foreign pronunciation, Fiji, first promulgated by Captain James Cook, that these islands are now known."*

---

[22] https://www.thecoconet.tv/coco-talanoa/pacific-blog/reviving-fijis-traditional-architecture/
[23] https://en.wikipedia.org/wiki/Fiji#Early_settlement

The Indigenous Architecture of Vanuatu

The House that can withstand a Cyclone[24]: how traditional dwellings are returning in Vanuatu. Since Cyclone Pam devastated Vanuatu, locals are returning to the *saeklon haos*, made from vines, palm fronds and grasses.

Figure 106 Traditional house construction on the island of Aneityum. Photograph: Gregory Plunkett

When Cyclone Pam devastated Vanuatu and concrete buildings collapsed, their iron roofing was blown away, and there was no loss of life in the traditionally built structures known as *saeklon haos* (cyclone house).

The vast majority of villagers still build their own homes from local materials. Most dwellings are traditional Melanesian houses with earth or coral floors, no glass windows, palm, bamboo, or cane walls and roofing. The most widely used exterior construction material was bush.

Figure 107 Left: Historical image from 1875 showing one of the earliest representations of the nakamal in Vanuatu; Middle: The Taloa Farea on Nguna Island post Tropical Cyclone Pam; Right: The Moriu Nakamal on Epi Island post Tropical Cyclone Pam

Traditional meeting places - Nakamals[25] are meeting places for Chiefs in Vanuatu and play a significant role in Vanuatu's kastom (custom/culture). The nakamal traditionally accommodates the functions of kastom governance, kastom court and other kastom ceremonies.

Vanuatu[26], officially the Republic of Vanuatu (French: République de Vanuatu; Bislama: Ripablik blong Vanuatu), is an island country in the South Pacific Ocean. The archipelago of volcanic origin is 1,750 km (1,090 mi) east of northern Australia, 540 km (340 mi) northeast of New Caledonia, east of New Guinea, southeast of the Solomon Islands, and west of Fiji.

---

[24] https://www.theguardian.com/world/2021/aug/09/the-house-that-can-withstand-a-cyclone-how-traditional-dwellings-are-making-a-comeback-in-vanuatu
[25] https://unesdoc.unesco.org/ark:/48223/pf0000248144
[26] https://en.wikipedia.org/wiki/Vanuatu

Chapter 7
Vernacular Architecture in Papua New Guinea

Figure 108 Left: Marketplace of Kalo, British New Guinea, 1885; Right: Lae International Hotel, 4th Street Lae Archway

Papua New Guinea[27] (abbreviated PNG; Tok Pisin: Papua Niugini; Hiri Motu: Papua Niu Gini), officially the Independent State of Papua New Guinea (Tok Pisin: Independen Stet bilong Papua Niugini; Hiri Motu: Independen Stet bilong Papua Niu Gini), is a country in Oceania that comprises the eastern half of the island of New Guinea and its offshore islands in Melanesia (a region of the southwestern Pacific Ocean north of Australia). Its capital, located along its Southeastern coast, is Port Moresby.

Etymology

"New Guinea" (Nueva Guinea) was the name coined by the Spanish explorer Yñigo Ortiz de Retez. In 1545, he noted the resemblance of the people to those he had earlier seen along the Guinea coast of Africa. The word papua is derived from an old local term of uncertain origin. Guinea, in its turn, is etymologically derived from the Portuguese word Guiné. The name is one of several toponyms sharing similar etymologies, ultimately meaning "land of the blacks" or similar meanings about the dark skin of the inhabitants.

European encounters

Little was known in Europe about the island until the 19th century, although Portuguese and Spanish explorers, such as Dom Jorge de Menezes and Yñigo Ortiz de Retez, had encountered it as early as the 16th century. Traders from Southeast Asia

Figure 109 Tribe in Papua New Guinea with wigwam at a settlement; Source: 1889 Geographie

---

[27] https://en.wikipedia.org/wiki/Papua_New_Guinea

visited New Guinea beginning 5,000 years ago to collect bird-of-paradise plumes.

Bee species with PNG type locality:

Austroplebeia cincta (Mocsáry in Friese, 1898)
Trigona cincta Friese 1898: 431: Holotype (HNHM, worker): examined, "Friedrich- / Wilh.-hafen", "N. Guinea / Biró 96" (taxonomy); Type locality: PAPUA NEW GUINEA "Neu-Guinea (Friedrich Wilhelmshafen)" (1 worker); Friese 1909("1908"): 354, 356 (distribution); INDONESIA, Irian Jaya/PAPUA NEW GUINEA, Cyclopen Gebirge; Moaif; Moso; Sentani; Tawarin; Timena; Friese 1909: 272, 273 (citation, distribution, key to species); INDONESIA, Irian Jaya/PAPUA NEW GUINEA, Cyclopen Gebirge; Moaif; Moso; Sentani: Tawarin; Timena;

Figure 110 Austroplebeia cincta (Mocsáry in Friese, 1898)

Tetragonula biroi (Friese, 1898)
Trigona birói Friese 1898: 428, 429: Syntypes (ZMHB, two workers): examined, "N. Guinea / Biró 96", "Friedrich- / Wilh.-hafen", "*Trigona / biro*i / det. Friese 1897" (1 worker); "Philippinen / Schadenberg", "*Trigona / biroi* / det. Friese 1897 / n.sp.", "Coll. / Friese" (1 worker) (key to species, taxonomy); Type locality: PHILIPPINES "Philippinen (Schadenburg)" (few workers); "N-Guinea (Berlinhafen und Friedrich Wilhelmshafen, 1896 Biró" (3 workers); The specimen was collected by Lajos Biro, a Hungarian and together with Friese an Austrian who had liberties in Northwestern PNG at a time when Germany colonised that region of New Guinea from 1884 to 1919.

Tetragonula clypearis (Friese, 1909)
Trigona laeviceps variety ***clypearis*** Friese 1909("1908"): 356, 358, fig. 15-3: Unknown; paralectotype
(ZMHB (1)) (distribution, taxonomy); Type locality: INDONESIA (IRIAN JAYA) "Manikion, 14.-28. February 1903"; "Wendesi 29. June 1903, New Guinea" (worker); Friese 1909: 275;

Tetragonula laeviceps (Smith, 1857)
Trigona læviceps Smith 1857: 51: Holotype (OUMNH). Moure (1961) indicated that the BMNH type specimen (17b.1184) came from Mt. Ophir, which was not the type locality, and the specimen cannot be considered a true type. Baker (1993) located three specimens in OUMNH and labelled one as the holotype. Unfortunately, that specimen is identical to *fuscobalteata,* according to Baker (1993), but this issue will be resolved in a separate paper with C.D. Michener. Citations below for *laeviceps* are broad, as species limits are uncertain and may include *valdezi* and *testaceitarsis*. These taxa have previously been proposed as junior synonymous of *laeviceps* but are slightly larger (taxonomy); **Type locality**: SINGAPORE "Singapore" (worker); Smith 1858: (propolis, wax); Smith 1859: 135 (distribution); INDONESIA, Aru; INDIA (misidentification); SINGAPORE; Smith 1865: 93 (distribution); "**NEW GUINEA**"; Parish 1866: (common name (pwai-ngyet), illustration, nest, propolis); Smith 1871: 395 (distribution); **PAPUA NEW GUINEA**, Sepik-Bivak; Ihering 1912: Irian Jaya/PAPUA NEW GUINEA, Etna Bai; Jasa; Merauke; Njao; Zoutbron; Cockerell 1918: 385 (comparative notes); PAPUA NEW GUINEA, Bisianumu, Sogeri Plateau; Bubia, vicinity of Lae; Kapagere, near Rigo; Koitakinumu Estate, Sogeri Plateau; Port Moresby; Michener 1965: 231 (citation);

Tetragonula sapiens (Cockerell, 1911)
Trigona sapiens Cockerell 1911: 176: Holotype (ANIC, worker) (comparative notes, taxonomy); **Type locality**: SOLOMON ISLANDS "Solomon Islands, July-August, 1909 (Froggatt)" (worker); Cockerell1929e: 242-243 (comparative note, distribution, taxonomy); SOLOMON ISLANDS, Lavoro Plantations, Guadalcanal Is.; New Georgia, W. Solomon; Cockerell 1936: 225 (distribution); SOLOMON ISLANDS, Halaita, Nggela; Schwarz 1939a: 111, 113 (synonymy); "NEW GUINEA"; PHILIPPINES; SOLOMON ISLANDS; Dollin 1997: (characteristics);

Platytrigona lamingtonia (Cockerell, 1929)
Trigona lamingtonia Cockerell 1929e: 243: Holotype (AMS, worker): examined, "Mt. Lamington / Northern Division / Papua v.1927 / C.T. McNamara", "Trigona / lamingtonia / Ckll TYPE", "HOLOTYPE", "Australian Museum K 238299". The planifrons species group (taxonomy); **Type locality**: PAPUA NEW GUINEA "Mt. Lamington, May 1927 (C.T. McNamara)" (workers);

Platytrigona planifrons (Smith, 1865)
Trigona planifrons Smith 1865: 93-94: Holotype (OUMNH, worker): According to Baker (1993), the specimen is labelled" 'N' [New Guinea (Allen); white disc] and '*Trigona planifrons* 'Smith' ". The BMNH "type" (17b.1180) indicated by Moure (1961) is not a true type (Baker, 1993). The *planifrons* species group (taxonomy); **Type locality**: INDONESIA **(IRIAN JAYA)** "New Guinea" (worker); Smith 1871: 396 (distribution); "NEW GUINEA"; Friese 1909("1908"): 354, 356, fig. 15-8 (distribution); INDONESIA, Irian Jaya/PAPUA NEW GUINEA, Digul Fluss; Etna Bai; Jamur Gebiet; Manikion; Manokwari; Merauke; Moaif; Tawarin; Friese 1909: 272, 273, 274 (distribution, key to species); INDONESIA, Irian

Figure 111 The National Emblem of PNG

Jaya/PAPUA NEW GUINEA, Digul Fluss; Etna Bai; Janour Gebiet; Manokwari; Merauke, Moaif; Tawarin; Friese 1915: 4 (distribution); INDONESIA, Irian Jaya/ PAPUA NEW GUINEA, Digul Fluss; Etna Bai; Manikion; Manokwari; Merauke; Moaif; Sattelberg; Sekofro; Simbang; Tawarin; Tjahe Fluss; Zoutbron; Cockerell 1923: 241 (key to species); Cockerell 1929e: 242 (comparative note, distribution); PAPUA NEW GUINEA, Fly River; Mt. Lamington; Schwarz 1948: 83, 109, 118; Moure 1961: 203, 204 (systematic position); Camargo 1988: 372 (distribution); PAPUA NEW GUINEA, N. Guinea, Simbang, Huon Golf;

Heterotrigona* (*Sahulotrigona*) *paradisaea [Etym: The specific epithet is based on the generic name *Paradisaea* Linnaeus (Aves: Passeriformes: Paradisaeidae), the genus of the emperor bird-of-paradise (*Paradisaea guilielmi* Cabanis), which is the national emblem of Papua New Guinea.] Note: The name is meant to honour the unique biota cultivated by the indigenous peoples of New Guinea.

Heterotrigona* (*Sahulotrigona*) *tricholoma Engel, new species [Etym.: The specific epithet is taken from the Greek words trichós (τρῐχός), meaning "hair", and lôma (λῶμᾰ), meaning "border" or "fringe", and is a reference to the fringe of setae bordering the mesoscutum posterolaterally.] ***Heterotrigona* (*Sahulotrigona*) *taraxis*** Engel, new species [Etym.: The specific epithet is taken from the Greek taraxis (τᾰρᾰχῆς), meaning "confusion", and is a reference to the historical confusion regarding the identity of this species.]
Comments: This species has been long confused as "*Trigona atricornis* Smith". A recent examination of the holotype of *T. atricornis* in the Oxford University Museum of Natural History reveals that Smith's (1865) species belongs to the genus ***Papuatrigona*** Michener & Sakagami, wherein it is a senior synonym of T. genalis Friese (vide infra) (Engel 2019).

Vernacular architecture of West Papua

Etymology: From Malay pepuah ("curly (of hair)") for the natives of the island of New Guinea; from Ternate/Tidore papo ua ("not united; not coalesced"), referring to the territory that geographically was far away from the Sultanate of Tidore (and thus not united).

Bevak is the traditional structure of the Kanume tribe of South Papua, using tree bark as a roofing material. The example below is used by a beekeeper (Theresia Agnesia Maturbongs, alias Hesty) in Yanggandur, Merauke. The roof is made from the skin of the Mbus Tree, which people here call the tree Mbus. History doesn't know its true name but it resembles the paperbark tree (*Melaleuca cajuputi* or *Melaleuca leucadendra*). Tetragonula cf mellipes and Austroplebeia cincta are bee colonies in cuboid wooden boxes.

The huts are like this in Merauke because the weather here is very hot, so there are no walls. The roof uses Melaleuca sp. tree bark because the 'paperbark' has a thick layer, is absorbent, and stores moisture to keep cool and damp in the hut (Pak La Hisa, Ranger in Wasur National Park, West Papua).

Figure 112 Bevak (traditional home of Kanume tribe South Papua) done by Theresia Agnesia Maturbongs aka Hesty and Madu Pokos Merauke— at Bevak Lebah Tetragonula Cf Mellipes _ Simpang Yanggandur Rawa Biru, Yanggandur, Merauke, Papua.

Bee species relevant to Papua:

Papuatrigona genalis (Friese, 1909)

Trigona genalis **Friese** 1909("1908"): 354, 356, 357, fig. 15-2: **Lectotype** (ZMHB, worker): examined, designated Michener 1990, "Manikion / 16.28.II.03", "Trigona / genalis / 1904 Friese det. / n. Fr.", "Type" (red label), "Coll. Friese", "Zool. Mus. / Berlin"; paralectotype (ZMHB (3)) (distribution, taxonomy); **Type locality: INDONESIA (IRIAN JAYA)** "Manikion, 14.-28. February 1903 (Neu-Guinea)"

Figure 113 Manokwari House of West Papua
Source: http://juliesartoni.blogspot.com/2012/03/traditional-architecture-of-indonesia.html

Manokwari

West Papua has more than 300 indigenous tribes that live nomadically or move where there are resources. The traditional house has only one form of "Honai House", though different tribes, languages and lifestyles. Honai houses are built of wood and straw, and building materials are readily available in natural surroundings. Manokwari is also the name of the capital city of West Papua province.

Home honai has a circular form with half of a coconut-shaped roof. Also, an open space interior with two levels, which on the top level is used as a bed. Honai house has one small door, usually without windows and ventilation, to be safe from beasts and keep the room temperature warm. In the centre of the room is a furnace used for cooking and heating. One Honai house consists of several families or single heads of households with multiple wives and children and is a gathering place for families.

The Asmat People.

Figure 114 Asmat on the Lorentz River, photographed during the third South New Guinea expedition in 1912–13.

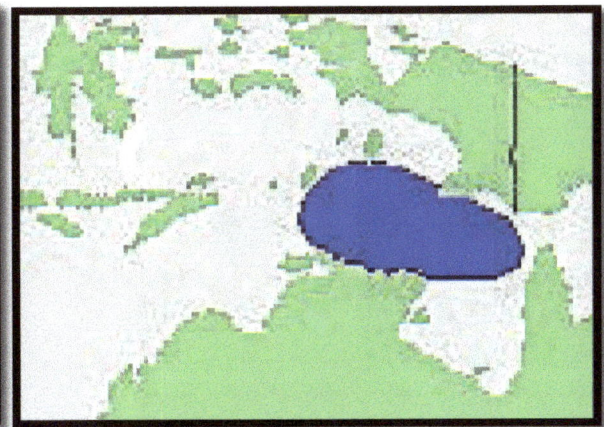

Map 3 Arafura Sea Edited by M. Minderhoud - https://ms.wikipedia.org/wiki/Fail:Locatie_Arafurazee.PN

The Asmat[28] is an ethnic group of New Guinea residing in South Papua, Indonesia. The Asmat inhabits a region on the island's southwestern coast bordering the Arafura Sea (See Map 3)

Figure 118 is a scene depicting a traditional river mangrove or estuary village of the Asmat people. Due to the daily flooding in many parts of their land, Asmat dwellings have typically been built two or more meters above the ground, raised on wooden posts. Some with elaborate stylized wood carvings, such as the '*bisj*' pole, are designed to honour ancestors.

Figure 115 Traditional; River mangrove or estuary village of the Asmat people of Papua, Indonesia,

Figure 116 Asmat shields

Figure 117 Wood carving depicting Asmat Hut and daily life

---

[28] https://en.wikipedia.org/wiki/Asmat_people

Chapter 8

Vernacular architecture in Timor Leste and Nusa Tenggara Timor

Map 4 Vernacular architecture in Timor Leste, West Timor & Rote Island, NTT, Indonesia.

Figure 118 Vernacular dwelling structures in Benteng None Village, West Timor, NTT.

1. *Benteng None* – possibly the most interesting tribal village in West Timor where one gets culturally shocked – the head-hunter tribe still has its shaman conducting magic and animist rituals (located in the Gunung Mutis Area). This village (*Benteng None*) is known to have the fiercest warriors in Timor. They only stopped headhunting in the 1940s. Headhunting involved chopping off an enemy's head, having a ceremony with the head for four days, and presenting it to the king. The rocks surrounding this village are very clearly coral, even though we are in the mountains. Unlike volcanic Flores Island, Timor Island was created by pushing land up from the sea.

2. The *Fatumnasi* tribe was the most modern tribe of all four tribes near Kupang, although their conical hut construction varies from the southern to the northern regions in the mountains.

Figure 119 Left: Houses in Fatumnasi Highland region; Right Tribal house in Amarasi, Kupang Province

3. Tamkessi village – a scenic countryside in the north-central territory (access by scooter/local bus/ojek via Manufui. The village of Tamkessi is on a rocky hill, and all the paths are tricky and slippery due to the rocks. Therefore, it was quite stressful not to fall down and not drop anything. Luckily, they don't cut heads or sacrifice tourists because hardly anyone can escape from this remote location.

Figure 120 Left: Fatumnasi Rumah Bulat (Roundhouse); Right: Tamkessi Conical Hut in West Timor.

4. Today's settlement area of the Bunak people (See Figure 118 p.72) is located in the mountains of central Timor, ranging from the East Timorese town of *Maliana* in the north to the Timor Sea in the south, where both the Bunak and the Tetun communities often live side by side in coexistence.

Figure 121 A traditional house called laco, for meeting the families that are part of a Lulic (sacred) house and also to receive guests, circa 1968–1970.

5. The **Kemak** (Portuguese: Quémaque, also known as Ema) people are an ethnic group numbering 80,000 in north-central Timor Island. They primarily live in the Bobonaro, East Timor district, while the rest live in the East Nusa Tenggara province of Indonesia.

Before the colonial period, Atsabe was one of the centres of Timor under the *koronel bote* (meaning "Kings") of Atsabe Kemak, which dominated the entire Kemak inhabited areas in East Timor until the colonial period. The Kemak areas include the north of present-day Bobonaro, the northern Ainaro, and the Suai area tributary to Atsabe.

Earth Tremor Resistance Sacred House of Timor Leste.

Figure 122 Inspired by the Uma Lulik or Sacred House of Timor Leste.

Traditionally, in Timor-Leste, the *Uma Lulik* (Sacred House) is the centre, the umbilical cord between the past and the present. For the dead, it's a timeless place where history is written constantly. For those alive, it is a secured reservoir of memories and wisdom.

The Bunak (also known as Bunaq, Buna', Bunake) people are an ethnic group that live in the mountainous region of central Timor, split between the political boundary between West Timor, Indonesia, particularly in Lamaknen District and East Timor.

The *Uma Lulik* covers two aspects i.e., one is that it is a building made of only natural materials such as timbers, bamboo, wooden planks, twine, *Arenga pinnata* fibre rope and others. The other aspect is intangible in that the use includes ceremonies and rituals, the history of the house, and people's beliefs,

Figure 123 Left: Holy House in Maununo, Suco Cassa, Ainaro Subdistrict, Ainaro District, East Timor; Right: Man in Fatuc Laran, Lactos, Cova Lima District, East Timor, 2009. Source: https://en.wikipedia.org/wiki/Bunak_people

for example, as a protective house. The *Uma Lulik* also provides a scenario for adapting rocks and plants as the place for people to "communicate" with their ancestors[29].

Community life is centred around sacred houses (Uma Lulik), physical structures serving as a symbol and identifier for each community. The architectural style of these houses varies between different parts of the country. The house as a concept extends beyond the physical object to the surrounding community. Kinship systems exist within and between houses.[30]

Bee species record relevant to East Timor:

Tetrigona vidua (Lepeletier de Saint Fargeau, 1836)
Syn. Melipona vidua Lepeletier de Saint Fargeau 1836: 429: Syntype (MNHN, one worker): putative syntype examined, "Museum Paris / Bengale / Diard & Duvaucel 1815", "vidua", "Diard et Duvaucel", "type", "TYPE", "*M. vidua* / Lep. S. Farg / Bengale".

Figure 124 Sacred house (Lee Teinu) in Lospalos

I consider this a true type: Alfred Duvaucel (1793-1825) and Pierre-Médard Diard (1794-1863) arrived on their first expedition to Kolkata, India, in January 1818 and spent years collecting in the Indo-Malayan region. The label date is wrong as they were still in France in 1815 but did reside for long periods in India later (Claeys: 1954), and I consider the label provenance merely the port of embarkation of the specimen rather than the collecting locality. The type is similar to *melanoleuca* and could be considered a senior synonym. However, melanoleuca is a variable taxon and is here retained until further studies may document if it is only a single taxon (taxonomy); **Type locality**: INDONESIA/**E TIMOR** "Ile de Timor" (unknown);

---

[29] Sacred Houses in Timor-Leste: Traditional Architectural Knowledge and Practice. January 2012
In book: Local Knowledge of Timor! (pp.2-12) Publisher: Haburas Foundation – UNESCO Editors: Demetrio do Amaral de Carvalho

[30] https://en.wikipedia.org/wiki/East_Timor#Politics_and_government

Chapter 9

Vernacular Architecture in South China

The distribution area of stingless bees in China is limited and only distributed in tropical and subtropical regions like Yunnan, Hainan, Guangxi, Tibet and Taiwan Provinces. (Zheng, 2022) Some of their nests were found in rocks of limestone forests and bamboo logs, while *L. flavibasis* nests were also found in telegraph poles and brick buildings. (Y.-R. Li et al. 2021)

Reminiscing my time in Kunming more than two decades ago, I did not see any stingless bees then, but I was intrigued by the Muslim culture there and their vernacular architecture.

Figure 126 Vernacular architecture with a typical Chinese courtyard; bottom photos are AHJ with local folk in a Muslim restaurant in Kunming City, Yunan

Figure 125 Wooden houses in an old Chinese village in Yunnan, China

Figure 127 Dai Theravada Buddhist temple in Menghai County, Xishuangbanna. https://en.wikipedia.org/wiki/Yunnan#/media/File:Weihan_Manduan_Temple_Menghai.jpg

Figure 128 Tuogu Mosque in Ludian County. https://en.wikipedia.org/wiki/Yunnan#/media/File:%E6%8B%96%E5%A7%91%E6%B8%85%E7%9C%9F%E5%AF%BA_-_panoramio_-_hilloo_(50).jpg

We will notice great differences between architectural styles between suburban and rural village designs. The three major different religious faiths also follow varied styles in their buildings. Interestingly, anthropologists speculate that the origins of Malay people are from Yunnan. Looking at rural dwellings, one can sense the Malayness they exude.

Figure 129 Traditional Dai village house in Jinghong, Yunnan A traditional house in a Dai village near Jinghong, with living space on the top, and storage space below.

Figure 130 Traditional Dai village house in Jinghong, Yunnan Traditional, rustic, wooden hut house in a rural village with living space on the top, and storage & work space below

Figure 131 Manyangguang Dai Ethnic Village in Jinghong City, XishuangBanna

As we get to the suburban houses, the hip & gable roof style is all but familiarly Malay houses with finials like the Malay 'tunjuk langit' spires and very Kelantanese pediment. The entrance stairway is similar to the Langkawi style under the porch roof.

The Origin of Malay Dwelling and the Yunnan Theory

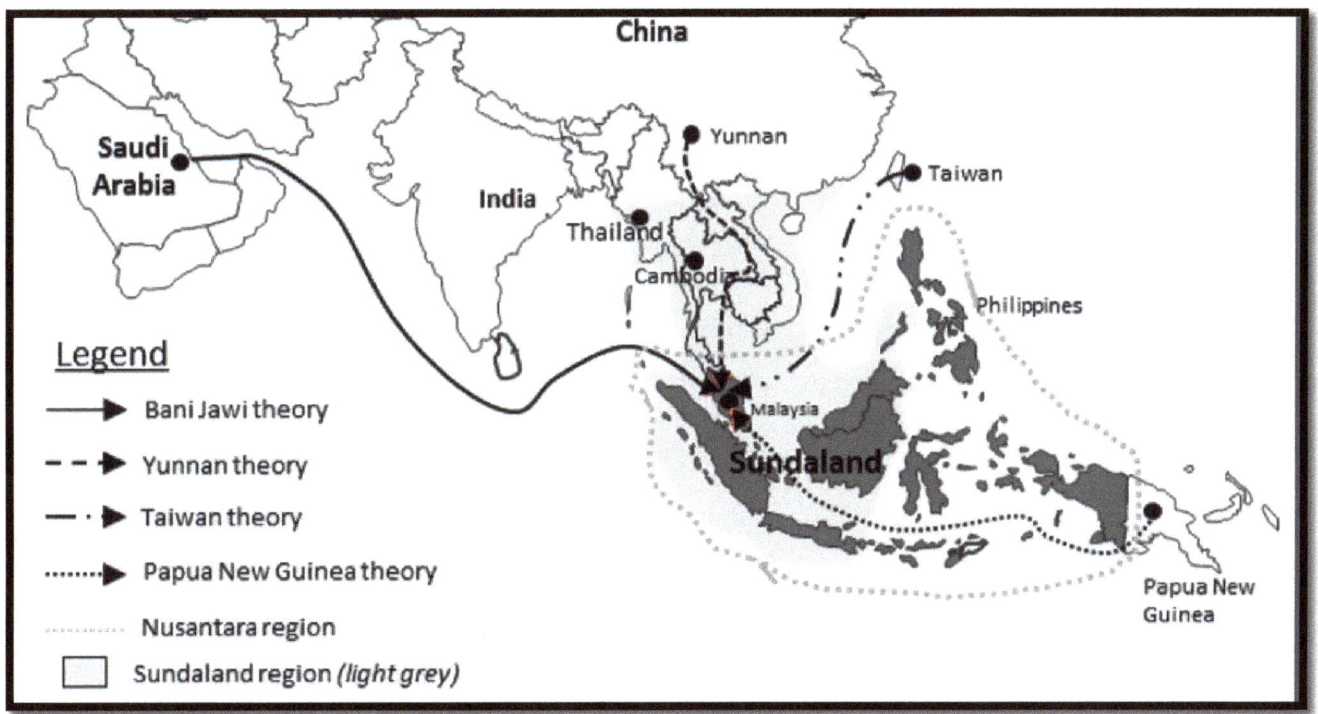

Map 5 Migration pattern showing the origin of Malays according to developed theories.
(Redrawn from Shalini Parthipan & Seri Mirianti Ishar (2022)

Figure 132 Replica designs in an effort to preserve the Malay culture and architecture in Peninsula Malaysia

Exploring the Yunnan theory, we look at the vernacular architecture of rural Yunnan, which has a semblance of rural Malay architecture. The stilts of the *Rumah Panggung* and the Hip & Gable roof style are very similar.

We shall explore dwellings on the Taiwan theory trail in the next sections.

Vernacular architecture on Hainan Island, China

The tribal Proto-Malays are believed to have migrated from Yunnan, China, about 4,000–6,000 years ago. They were once probably people of coastal Borneo who expanded into Sumatra and the Malay Peninsula as a result of their seafaring way of life (Hatin, 2011). This route from Yunnan must be via Hainan Island, where houses are roofed with upturned boats.

According to some scholars[31], Hainan was originally attached to the Northeastern part of Vietnam; however, the island was formed after it physically broke away from Vietnam due to a volcanic eruption and drifted southeast near China after the Mesozoic millions of years ago.

The Baiyue people are among the earliest Kra-Dai residents to arrive on Hainan Island. They are believed to have settled there at least 2 to 6 thousand years ago and carry genetic markers from ancient people who reached the island between 7 and 27 thousand years ago[32].

Figure 135 Hainan's Maona Village promotes rural tourism to increase locals' income and expedite rural revitalization Source: XinhuaEditor: huaxia2022-04-12 21:58:47

The word "pile-dwelling" was first seen in "Wei Shu Liao Zhuan", in which "building blocks are built according to trees to occupy them, which is called "Pile-dwelling". It can be seen that pile-dwelling refers to the houses built on the piles. Pile-dwelling of the Li nationality in Hainan are living fossils in the history of Chinese architecture.

Figure 134 Auxiliary pile-dwelling of Hainan Li Nationality

Figure 133 Wooden Houses with cane roofs at Hainan Island — Photo by ADavydova

---

[31] Hatin WI, Nur-Shafawati AR, Zahri MK, Xu S, Jin L, Tan SG, Rizman-Idid M, Zilfalil BA; HUGO Pan-Asian SNP Consortium. Population genetic structure of peninsular Malaysia Malay sub-ethnic groups. PLoS One. 2011 Apr 5;6(4):e18312. doi: 10.1371/journal.pone.0018312. PMID: 21483678; PMCID: PMC3071720
[32] https://en.wikipedia.org/wiki/Hainan

"Native bee species include the Hainan Honey Bee (Apis cerana hainanensis) and two species of stingless bee (Tetragonula pagdeni and Lepidotrigona ventralis). The Hainan Honey Bee is the major pollinator of many native plants, including valuable herbs like the cardamom (Alpinia oxyphylla). Therefore is important to the local ecosystem. As for stingless bees, their honey is known for its unique flavour and commands a higher price in the market than honeybees[33]." (Padilla, H. & Li, J. 2018)

Figure 136 Left: Lepidotrigona ventralis; Right: Apis hainanensis

Blue-tailed bee-eaters[34] seen in Haikou as reported in Xinhua News (Published: Apr 18, 2022[35]). With these new threats to the rising Meliponiculture activity, vernacular architecture can play a protective role from the influx of Bee-eaters on the island. The overhang of cane leaf roofs may pose some protection from these predators.

Figure 137 Blue-tailed bee-eaters

---

[33] https://www.kfbg.org/en/featured-projects-in-mainland-china/eco-beekeeping
[34] https://en.wikipedia.org/wiki/Blue-tailed_bee-eater#/media/File:Blue-tailed_Bee-Eaters_with_Dragonfly.jpg
[35] https://www.globaltimes.cn/page/202204/1259568.shtml

Lepidotrigona ventralis hoozana in Taiwan:

Photos by Dr I-Hsin Sung, Assistant Professor, Department of Plant Medicine, College of Agriculture, National Chiayi University. E-mail: ihsinsung@mail.ncyu.edu.tw

Figure 138 Right: Taiwanese meliponiculture of Lepidotrigona ventralis hoozana; Left: Nest entrance,

Interview with Dr. I-Hsin Sung by AHJ: "Good Day to you, sir. I am making some notes on the Meliponiculture of Ind-Malayan Stingless Bees in The ASEAN region. I hope to get some information about the culture of *Lepipdotrigona hoozana* in Taiwan. Can you tell me if any Bee farm in Taiwan cultures *L. hoozana* and whether any honey harvesting is carried out? Are they purely keeping bees for pollination? If so, for what crops are they used in pollination? What are the yield improvements as compared to non-SB pollination? Any info on the location and altitude will also be valuable. Are stingless bee hives or products from the SB hives commercially transacted in Taiwan?"

Dr I-Hsin Sung: The stingless bee *Lep. hoozana* so far in Taiwan is still rare, maybe distributed in deep mountains. Only a few people can get nests. I know few people rearing nests, but they are for a hobby. I have written one paper, "Geographic Distribution and Nesting Sites of the Taiwanese Stingless Bee *Lepidotrigona ventralis hoozana* and an Unidentified Subspecies of Asian Honeybee *Apis cerana* in Taiwan (Hymenoptera: Apidae)". Unfortunately, it was written in Japanese.

There is insufficient data on whether this lone representative of Indo-Malayan stingless bees in Taiwan is kept for honey or other products; however, some pollination activities have been recorded.

These bees (together with honeybees) pollinate the following crops in Taiwan: Anacardiaceae (mango), Compositae (sunflower), Cruciferae (oilseed rape), Cucurbitaceae (cucumbers, melon, sponge cucumber, watermelon), Myrtaceae (guava), Orchidaceae, Rosaceae (berries, pear, plum), Sapindaceae (litchi, longan), Rutaceae (citrus fruits) (Sung, 2010)

Lepidotrigona ventralis hoozana (see Figure 118) has been confirmed in four counties: Taichung, Nantou, Chaiyi and Kaohsiung. Its distribution range was confined to submontane and mountainous areas at 250-2,500 m altitude, where primary and secondary forests remain. A total of 40 stingless bee nests, including 33 domesticated ones, naturally occurred in the hollow cavities of trees. The lack of stingless bees in the lowlands could be due to a deficiency of suitable host trees. (Sung, Yamane, Ho, & Chen, 2006)

Figure 139 Lepidotrigona ventralis hoozana

Song Yixin author Creative Commons license. Information provided Taiwan Encyclopedia of Life Description : *Lepidotrigona ventralis hoozana* Middle Name: Taiwan stingless bee, Taiwan needle bee, bee flies. Length:4.4-7 .8 mm. The main identifying features: Small species. Body black to dark brown, milky white belly section 1, in the chest around the periphery of the white bushy-haired, veins degradation, non-sting. Conservation: Rasmussen (2008) to this species and Southeast Asian sister species separated into separate species, which are endemic to Taiwan, the number of species on the already extremely rare, ecologically low degree of competition for resources with native wild phenomenon Toyo bees, and there a small amount of hunting pressure, conservation units worth attention. Distribution: very narrow, low to medium altitude distribution. Habitat: Nest in tree cavities.

*Lepidotrigona **hoozana*** (Strand, 1913), comb. nov. also [*Trigona ventralis hoozana* Strand 1913b: Type locality: Taiwan Hoozan][#] (Rasmussen, 2008) = Taiwanese racial name given by Strand 1913 – Hoozan = Fengshan[36], Kaohsiung County Taiwan.

---

[36] Fengshan city a.k.a. Hōng-soaⁿ in Taiwan Amoy Hokkien dialect, in Kaohsiung County.

Vernacular Architecture in Taiwan

Figure 141 An observational painting of a stilt house of one of the plains indigenous peoples as depicted in Liu Shi Qi's Taiwan Panorama Prints (六十七兩采風圖合卷).

Figure 140 A stilt house of the Truku people. (Author unknown, Public Domain Creative commons https://en.wikipedia.org/wiki/Architecture_of_Taiwan#/media/File:Mokka_and_their_house.jpg

The stilt house[37] of old Taiwanese indigenous people has an uncanny similarity with houses in Nusantara. The thatched Gable roofs with inward slants under the eaves. The ramp entrance resembles the bee house in a Javanese Apiary in Prawita Garden Farm in Banyumas, Central Java. The Mokka House is similar to a Malay rural village house.

Figure 143 A bee house built in Prawita Garden Apiary in Banyumas, Java, Indonesia

Figure 142 Typical Malay Rural Village House 1960s

---

[37] https://en.wikipedia.org/wiki/Architecture_of_Taiwan

Traditional Houses of Taiwan Tribal People

Figure 145 Left: Traditional bamboo house with a low-pitched roof of the Atayal people; Middle: Thatched roof Houses of the Seediq people in Paran Tribe (巴蘭部落); Right: Conical Houses of the Bunun people

Figure 144 Left: View of a Tao tribal Village; Middle: The Kakita'an ancestral house of the Amis people in the Tafalong Tribe (太巴塱部落); Right Stiled-round-ended house of the Puyuma people

Replica of Tribal houses and artefacts in the Formosan Aboriginal Culture Village[38]

Figure 146 Top Left: An imitation animal bone hut of the Tsou people; Top Middle: An imitation granary of the Rukai people; Top Right: An imitation of a ceremonial rack of skulls of the Paiwan people in Formosan Aboriginal Culture Village; Bottom Left: Totem Poles; Bottom Right: The Naruwan Theatre the in Formosan Aboriginal Culture Village

---

[38] https://en.wikipedia.org/wiki/Formosan_Aboriginal_Culture_Village It is in Yuchi Township, Nantou County, Taiwan

References for Stingless Bees in China & Taiwan

Engel, M. S., et al. (2022). A new genus of minute stingless bees from Southeast Asia (Hymenoptera, Apidae). *ZooKeys 1089: 53–72 (2022)*, doi: 10.3897/zookeys.1089.78000.

Hainan Normal University. (2019). The Primitive Pile-dwelling of Li Nationality in Hainan and Transition Xiong Hongli. *5th International Conference on Economics, Management and Humanities Science (ECOMHS 2019)* (p. DOI: 10.25236/ecomhs.2019.166). Haikou, Hainan 571158, China: Francis Academic Press, UK.

Kadoorie Farm. (n.d.). *Conserving the Hainan Native Bees and Stingless Bees*. Retrieved from SUSTAINABLE LIVING AND FARMING - FEATURED PROJECTS IN MAINLAND CHINA - ECO BEEKEEPING: https://www.kfbg.org/en/featured-projects-in-mainland-china/eco-beekeeping

Padilla, H. & Li, J. (2018). *Rubber-Based Agroforestry System in South China: Gaining Ground with Farmers.* Kadoorie Conservation China, Kadoorie Farm and Botanic Garden, Hong Kong SAR, China.

Parthipan, S. & Ishar, S. M. (2022). Perspectives On Ancestral Lineages And Genetic Markers Of Malay Population In Peninsular Malaysia. Jurnal Sains Kesihatan Malaysia 20 (2) 2022: 83-95, DOI : http://dx.doi.org/10.17576/JSKM-2022-2001-08.

Qu, Y., et al. (2022). *The newly rising meliponiculture and research on stingless bees in China–a mini-review*. Retrieved from Journal of Apicultural Research vol 61 issue 5: https://doi.org/10.1080/00218839.2022.2104568

Renping, W. & Zhenyu, C. (2006). An ecological assessment of the vernacular architecture and its embodied energy in Yunnan, China. *Elsevier Building and Environment Volume 41, Issue 5, May 2006, Pages 687-697*, https://doi.org/10.1016/j.buildenv.2005.02.023.

Sakagami, S. F., & Yamane, S. (1984). Notes on taxonomy and Nest Architecture of the Taiwanese stingless bee Trigina (Lepidotrigona) ventralis hoozana. *Ibaraki University*.

Starr, C. K., & Sakagami, S. F. (1987). An Extraordinary Concentration of Stingless Bee in the Philippines, with Notes on Nest Architecture (Hymenoptera: Apidae: Trigona spp). *Insectes Sociaux 34(2):96-107*.

Sung, I.-H. (2010). *Wild bees as crop pollinators in Taiwan.* Miaoli 363, Taiwan ROC: Sericulture and Apiculture Section, Miaoli District Agricultural Research and Extension Station, COA. Retrieved from http://www.niaes.affrc.go.jp/sinfo/sympo/h22/1109/paper_09.pdf

Sung, Yamane, S., Ho, K.-K., & Chen, W.-S. (2006, July 25). Geographic Distribution and Nesting Sites of the Taiwanese Trigona ventralis hoozana and an Unidentified Subspecies of Asian Honeybee Apis cerana in Taiwan (Hymenoptera: Apidae). *The Entomological Society of Japan, Jpn. J. Ent. (N.S.), 9(2)*, 34-35.

Sung, Yamane, S., Ho, K.-K., Wu, W.-J., & Chen, Y.-W. (2004, Sept). Morphological caste and sex differences in the Taiwanese stingless bee Trigona ventralis hoozana (Hymenoptera: Apidae). *Entomological Science, 7*, 263–269. doi:10.1111/j.1479-8298.2004.0

Sung, Yamane, S., Lu, S.-S., & Ho, K.-K. (2011). Climatological Influences on the Flight Activity of Stingless Bees (Lepidotrigona hoozana) and Honeybees (Apis cerana) in Taiwan (Hymenoptera, Apidae). *Sociobiology 58(3) January 2011*, 835-850.

Wang, L. & Yang, Z. (2012). Evaluation of sustainable vernacular architecture and settlements in Yunnan. *The Sustainable City VII, Vol. 2 905 (WIT Transactions on Ecology and The Environment, Vol 155, © 2012 WIT Press)*, doi:10.2495/SC120762.

Y.-R. Li et al. (2021). Species diversity, morphometrics, and nesting biology of Chinese stingless bees (Hymenoptera, Apidae, Meliponini). *Apidologie Original article © INRAE, DIB and Springer-Verlag France SAS, part of Springer Nature, 2021*, htttps://doi.org/10.1007/s13592-021-00899-x.

Y.-R. Li et al. (2022). *A new record of sweat-sucking stingless bee, Lisotrigona carpenteri Engel 2000, from a natural savanna in southwest China*. Retrieved from Journal of Apicultural Research vol 61 issue 5: https://doi.org/10.1080/00218839.2021.1987741

ZHENG, X., et al. (2022). *Recent Progress of Stingless Bee Honey.* doi 10.13386/j.issn1002-0306.2020120.285: Science and Technology of Food Industry, 2022, 43(1): 458–465. .

VOLUME 3, PART 3
~ ADDRESSING CALAMITIES & VERNACULAR ELEMENTS ~

Introduction to Part 3

By Abu Hassan Jalil

Indonesia is the country of the Ring of Fire volcanic belt and holds about 40% of the world's geothermal reserves. More than 200 volcanoes are located along Sumatra, Java, Bali and the islands of the eastern part of Indonesia, known as The Ring of Fire.

There are 53 active volcanoes in the Philippines. The Philippines belongs to the Pacific Ring of Fire, where the oceanic Philippine plate and several smaller microplates are subducting along the Philippine Trench to the East and the Luzon, Sulu and several other small Trenches to the West.

In the Pacific 'Ring of Fire', Indonesia and the Philippines are frequently bombarded with tremors and earthquakes. That which has proven to withstand such quakes through the test of time. Addressing the earthquake dilemma, one may explore ideas from island native architecture in the Nias Islands off Sumatra and the Mindanao tribes of Maranao at Lanao Lake.

Witnessing much devastation in some islands like Lombok and Luzon during travels in recent years. The most recent earthquake reports were included up to early March 2022 when writing this book. The ancient ideas and construction ingenuity are worth considering for new Bee Housing within the "Ring of Fire", which would help reduce the tremor-related damage that may occur.

Other predicaments are disaster mitigation measures in flood-prone areas necessary when it involves sanctuaries or repositories of bees. Learning from past disasters and going the extra mile, one may design floatable repositories or research units and protect them from extreme weather. Even thunderstorms and unpredictable damage like solar flares are taken into account.

Further, we may explore traditional houses on stilts, such as the traditional Lotud house in Sabah and others in the ASEAN region, typhoons, and strong gale-resistant structures like the Ivatan Bahay na Bato or stone houses in Batanes Island in North Philippines. The Ivatan houses have been typhoon and strong gale-resistant over centuries.

This Part also explores vernacular island architecture in Timor Leste, Nusa Tenggara Timor and Indonesian Papua on the Sahul shelf.

Chapter 10

Addressing the Earthquake Dilemma

At the time of writing, a string of earthquakes occurred in Central and North Sumatra. Started with a 6.2 magnitude on 25 Feb 2022 and then followed by consecutive tremors from 3rd Mar to 6th March

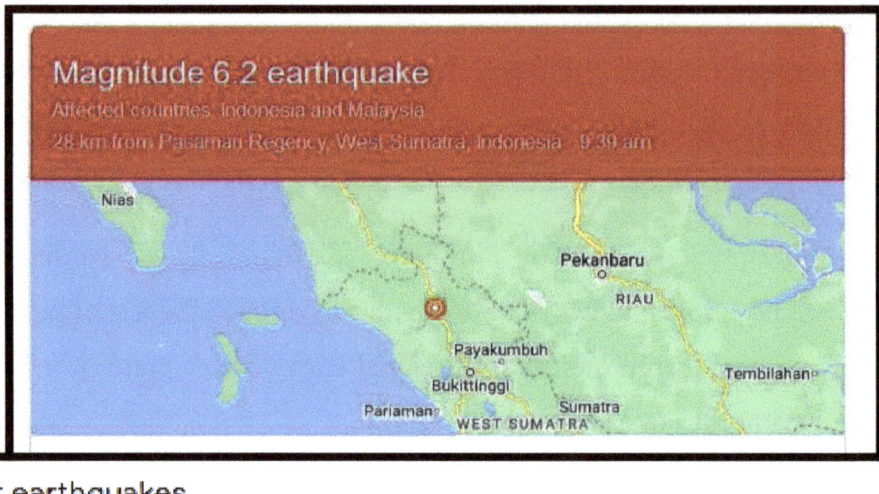
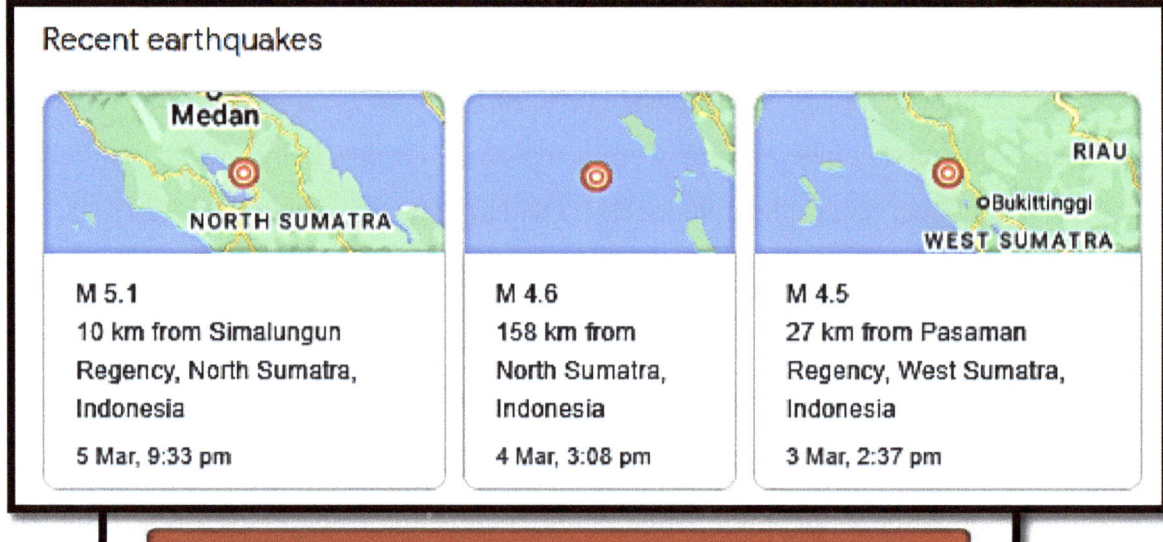

Figure 147 Earthquake Data 25 Feb to 5Mar 2022 in Sumatra

(Data source: https://earthquake.usgs.gov/earthquakes/eventpage/us6000h2an/executive)

Most recent Earthquake reports

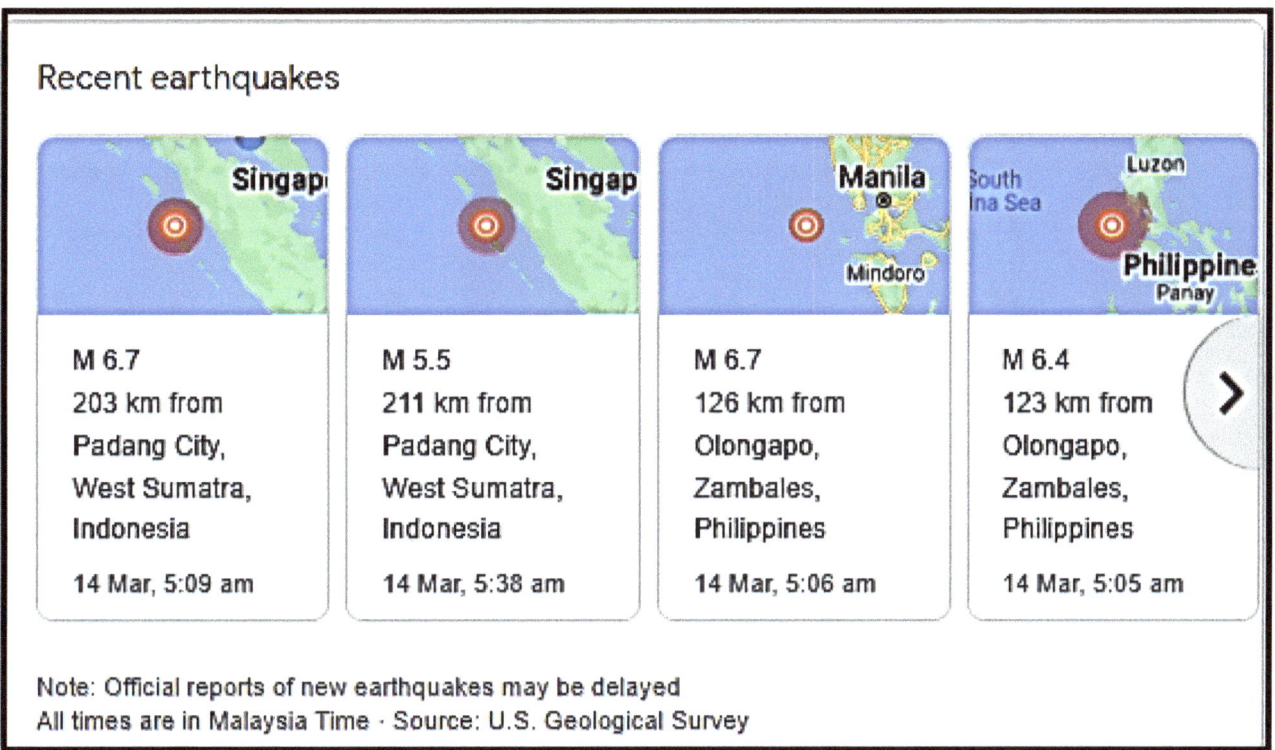

Figure 148 Recent earthquake in W. Java 22 Nov. 2022

In exploring and studying the various vernacular architecture, we find some literature regarding indigenous construction that reinforces house stability during earth tremors. This concept has prompted a reflection on Island vernacular architecture practised in areas prone to earthquakes and tremors. The Maranao community in Marawi, Mindanao, Philippines, practises a unique earthquake resistance technology. The posts are large tree trunks on rocks (about 2x palm palm-sized roundish -not rounded). One can imagine that during the tremors, the rocks act like castors.

Figure 149 Recent earthquakes in Tual, Maluku Islands as of 22 Nov. 2022 data.

Maranao Technology (contd.) "Once upon a time, when they built a traditional house, they would place stones under the pillars of the house, which consisted of large tree trunks. These stones serve as castors for the pillars of the house, and they can absorb the tremors of earthquakes that cause their houses not to collapse and remain strong."

"Many experts from outside have researched this technology, and among them are experts from Japan, where it is said that such technology had been further developed in temples, e.g., in Kyoto. The traditional Maranao houses are called Torogan; today, only one more original Torogan house remains. The Torogan house being visited is the only one that still exists, and it is over 300 years old." Malay Text: Hasanuddin Yusof; Translated by AHJ

Traditional Musalaki House

Figure 150 Traditional Musalaki House

Musalaki House is a typical residence of the **Ende Lio** Tribe in East Nusa Tenggara. The name is Musalaki and is taken from the Ende Lio language, namely *mosa,* which means head and *laki,* which means man. When combined, Musalaki means traditional leader or tribal leader. Quoting from the "Mobile-Based 3D Application Design for Traditional Houses in

Indonesia (2017)" by Andi Riskal Ir Andi Bolle, Musalaki House is included as a stilt house. The lower part of this house has a long hall used to receive guests. The main uniqueness of the Musalaki House is that its foundation is placed on a large rock, so it is not planted in the ground. This boulder is formed naturally and is usually oval and mounted vertically. The main benefit of using this foundation is to minimize cracks in the building in the event of an earthquake or other disaster. This house uses plank slats arranged parallel in one direction for the floor structure. In the preparation of the boards, they are deliberately made to have different heights for the entry and exit of air.

The main material for making the roof of the Musalaki House comes from straw. Uniquely, this straw is arranged and rests on the roof frame to avoid falling or being easily damaged. In addition, the roof of this traditional house also towers upwards. This towering symbolizes oneness with God Almighty as the creator. At first glance, if you pay attention, this traditional house's roof resembles a sailboat's shape. According to local people's beliefs, the ancestors of Ende Lio came by boat, so the roof of this traditional house was made as close as possible to a sailboat. According to the Ministry of Education and Culture (Kemdikbud), on the roof of the Musalaki House, you can find two ornaments that have important symbols, namely the kolo Musalaki (the head of the keda house) and the kolo ria (the head of the big house). It is believed that there is a spiritual connection between the two.

In summary[39], the Musalaki House has four unique features: It is only occupied by regional leaders, such as tribal chiefs, lurah or sub-district heads. As the name implies, namely musalaki, this traditional house is only intended to be a residence for tribal chiefs, *lurah* and sub-district heads. The roof is made of thatch. The roof of this traditional house is made of thatch, which is arranged regularly and rests on the roof frame. The foundation is mounted on a large oval stone, which is installed vertically. The pillars of the Musalaki House are mounted on a large and vertical oval stone so that the pillar is not stuck in the ground but is on the rock. The floor is made of plank slats arranged parallel and pointing in one direction. The floor of this traditional house is made of plank slats arranged parallel and pointing in one direction. The use of slats on this board is intended so that the humidity level on the house floor can be reduced.

---

[39] Source in Indonesian Language, "Artikel ini telah tayang di Kompas.com dengan judul - Keunikan Rumah Musalaki, Adat Nusa Tenggara Timur", Klik untuk baca: (https://www.kompas.com/skola/read/2021/05/05/135401269/keunikan-rumah-musalaki-adat-nusa-tenggara-timur?page=all)
Penulis : Vanya Karunia Mulia Putri; Editor : Serafica Gischa
Download aplikasi Kompas.com untuk akses berita lebih mudah dan cepat:
Android: https://bit.ly/3g85pkA; iOS: https://apple.co/3hXWJ0L

Traditional Nias Island House

Many researchers agree that Nias traditional houses (Omo Hada) are Asia's finest examples of vernacular architecture. Nias houses are elevated from the ground and built for defence, as Nias villages lived in perpetual conflict. They are built without nails and can withstand powerful earthquakes far better than modern houses.

Figure 151 Traditional Nias Island House

The strut braces and the posts are huge, heavy logs, making the whole house structure sturdy. Nias Island sits at the Sunda Trench and is part of the "Pacific Ring of Fire". It frequently experiences tremors before, during and after earthquakes.

Chapter 11

Disaster precautionary measures in flood-prone areas
By Abu Hassan Jalil

After the earth's tremors in March 2022, there was flooding on the East coast of Malaysia. Flooding was a yearly affair in this region, and in NE of Peninsula Malaysia, they sometimes even have water festivals (*Pesta Air*) during the end-of-year Monsoon season. However, it came late this year, and the whole stingless bee Repository in Sekayu, Terengganu, was immersed.

The flooding situation was detailed in the Blueprint for a 10-year development plan for the Malaysian stingless bee industry in 2019.

Figure 152 Proposed Hive Placement Pedestal in flood-prone areas

We've seen how the beekeepers tend to their colonies in swamp areas in canoes in the Amazon. However, this year's disaster was much worse. Last week, a large area in Kuala Lumpur was flooded, and many businesses and traffic came to a standstill.

With much concern for the MGVI Repository (although currently on much higher ground), the past couple of days caused the news[40] headlines to read Landslide in Ampang, Kuala Lumpur, caused damage to many homes and five buried alive found dead.

The Langkawi Malay Coastal Houses on stilts are vulnerable to tides, and the concrete stairway entrance anchoring the whole platform can be applied to bee farms in flood-prone areas.

---

[40] https://www.thestar.com.my/news/nation/2022/03/11/forty-eight-homes-in-taman-bukit-permai-told-to-vacate-area-declared-disaster-zone

Figure 153 Inspired by Langkawi coastal house on stilts.

We created a futuristic illustration for research stations, learning institutions, Cooperatives, beekeeper associations and future Bee repositories.

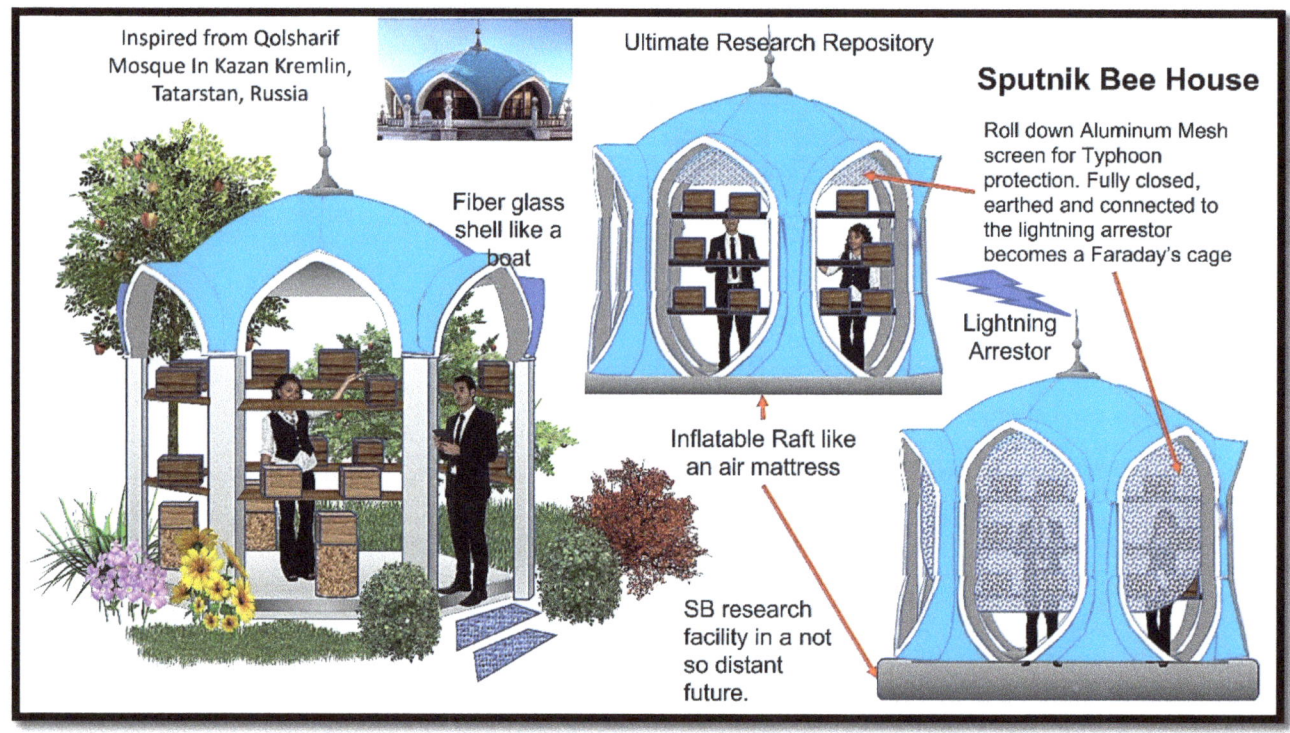

Figure 154 Inspired by Qolsharif Mosque in Kazan Kremlin, Tatarstan, Russia

Figure 155 Floating on floodwaters

The plan proposes an air mattress as an inflatable floor, inflated when necessary or when an imminent flood is cautioned. The design is for a sophisticated repository... for village breeders, you can use a permanent bamboo raft that will float once it's flooded. The lightning arrestor connects to the roll-down aluminium mesh to protect from lightning strikes. Fully closed, it converts into a 'Faraday's Cage'. The 'Faraday's Cage' is made from a conductive material so its electrons can move freely. For 'Faraday's Cage', the electric field exists on the exterior but not in the interior, so that's why the person inside doesn't get electrocuted.

A friend, Edhy Yanshah, in West Papua, hoping to set up his bee sanctuary on an island in Raja Ampat, Indonesia, is planning to take up this idea and use a bamboo raft instead of an air mattress.

The Korowai Tree House.

Figure 156 Wooden A-Frame Tree House Bee Shed on stilts *Figure 157 Inspired by the Korowai Treehouse*

Traditionally built on 20ft stilts to keep away from wild animals, enemies and floods, a modification for the convenience of Beekeeping is elevated from flood levels.

The Korowai are historically a cannibalistic tribe and have a way of life akin to the prehistoric Stone Age. Weapons and tools are stone wedges. The structure is built from trunk branches and twigs. Minimal walls or rails and no stairway but just a vertical notch ladder,

Chapter 12

Introduction to Vernacular Elements and Architectural Fusion
By Abu Hassan Jalil

Figure 158 Vernacular Fusion – Tanimbar Is. Dwelling & Tambi House of C. Sulawesi

This juncture brings us closer to a bit more modern construction with the advent of green roofs. These green roofs seem to provide better foraging opportunities. With fierce competition from other insects, the bees need to be housed in a structure that gives them the edge regarding resource partitioning. I also look at the possibilities of building floating structures that can be safe in flood-prone areas. This necessity came to light when recent monsoon weather extremes destroyed many Meliponaries. Examples of green-roofed boathouse ideas are explored.

This part examines Conical roofs and Dome Roofs on roundhouses[41] and why they are preferred for withstanding strong winds. It would only be natural to look at cultures that dwell in Teepees, Wigwams, Yurts and the like. From there, we identify some ethnic hanging hives and bee host containers. With these elements identified, one can then design structures that can accommodate the hanging beehives and plan for roof eaves of balconies for easy and convenient access to these hanging hives. The hanging hives can also be accommodated in bee racks and swing shelves.

---

[41] Editor's note: Yucatan, where I worked over 30 years, has oval houses with thatch roofs. Oval is key, I think.

Besides all that hiving and racks, we look into the pollination services involving Meliponines. Moving portable hives from farm to farm for a pollination service may harm the bees in the long run. What more, if the crops are located far from the Meliponary. Some ideas inspired by model caravans, roadshow buggies, trailers and carts that beasts of burden or mobile vehicles can move.

Ultimately, inspired by religious structures, we look at Islamic Structures and Mosques, a faith I am not alien to. There are archaic mosques with historical backgrounds, 'Onion' dome prayer halls and modernist designs of mosques worldwide. The Islamic tradition is synonymous with bees as a full verse is dedicated to An-Nahl (the bee in Arabic) in the Holy Quran (Surah 16: verses 68 & 69). Much of the Hadith and Sunnah also have literature on the medical benefits of Honey.

Ten elements of Vernacular Architecture that are suitable for Meliponiculture and bee housing.
1. Roundhouse or multi-sided house that will deflect strong winds. It would include conical or domed roofs on round, hexagon or octagon structures.
2. Mud houses that do not retain tropical heat waves. The mud bricks may derive from African or Indian ancient techniques and some Yunnan cultures.
3. Woven bamboo slats, tree bark or planks for wall cladding provide passive ventilation, thus unlike cement and concrete walls that retain heat and delay dissipation (undesirable in a tropical climate) except for rubble-walled houses, very stable in Typhoon prone areas.
4. Roof structures & shapes with organic material, e.g., palm leaves, wood bark or grasses, provide better heat dissipation, although some cultures utilize slate roof tiles.
5. Stabilized construction to withstand tremors and quakes, e.g., log posts on rock castors and log strut bracing. The Maranao of the Philippines and the Timorese construct houses with log posts on rock castors. Sumbawa Musalaki houses are built on embedded large boulders. Nias Island uses large logs as bracing and struts in its domicile construction.
6. Stilts, pedestals or raised platforms for flood damage mitigation. The cultures in coastal areas, along river banks, estuaries and wetlands, naturally build on raised floors.
7. Multi-tiered or storied for better bee foraging opportunities. Placing bee hives at higher levels allows the bees more access to tree canopies and taller vegetation.
8. Green roof possibilities for improved bee resource acquisition.
9. Sufficient courtyard or backyard space to accommodate Beescape. Having spacious courtyards allows for the bees' foraging segregation and control of pesticide exposure.
10. Indigenous or ethnic identity for Insect/Bee Tourism potential. A vernacular structure without an ethnic identity lacks Tourist attraction and, therefore, would lose out on Tourism potential.

Chapter 13

Conical roofs and Dome Roofs

Why are roundhouses hurricane-proof?

There is wind stability around roundhouses. Because the buildings are circular, the wind flows around them instead of putting pressure on one side of the home. Unlike a flat roof, a sloping roof deflects wind pressure, which can come loose during a hurricane[42].

Are roundhouses better for hurricanes?

A round or multiple-sided home is more resistant to hurricane-strength winds. The round design allows the wind to blow around the home, reducing pressure build-up on one side.

Why are roundhouses better?

With less exterior wall area, heating and cooling bills are lower. Cold winter winds flow smoothly around the house instead of leaking in and causing drafts. This circular design also withstands hurricane and tornado winds better. With the circular shape, each room can have two windows at an angle.

What about dome-shaped homes

Dome-shaped homes are the most energy-efficient since they have fewer corners. These domes allow wind to travel over the home easily without air pressure changes, which reduces air penetration and thus maintains a more even temperature. Cube-shaped homes are another good option.

Are round houses harder to build?

Roundhouses[43] can cost significantly less to construct. Traditional houses, with their multiple surfaces, are complicated structures. Round houses, though, are relatively simple. They use fewer materials and take less time to build.

---

[42] Editor's note: Tornadoes make houses blow up. The wind speed is high, but the pressure is extremely low- that is why we open windows on one side of the house. Same for hurricanes, I believe.
[43] A roundhouse is a type of house with a circular plan, usually with a conical roof. In the later part of the 20th century, modern designs of roundhouse eco-buildings were constructed with materials such as cob, cordwood or straw bale walls and reciprocal frame green roofs.

Traditional yurts are circular domed structures built by nomadic tribes in Central Asia. The largest yurt can be enhanced to withstand roof snow up to 100 pounds per square foot and wind up to 142 miles per hour.

It may be simple in description, a roof in the shape of a cone or a dome, but the variations to conical roofs are slightly complicated in terms of vernacular architecture. It can vary from an acute cone, parabolic, or hyperbolic to a dome hemisphere.

The vernacular conical roofs are mostly thatched with a multitude of natural fibres that are available to the native population of indigenous people. They came as assorted grasses like cogon grass, palm fronds, reeds, and straw (rice stalks). Some cultures use tree bark, animal skin or flat wooden slats (not sawn timber). Some use unusual materials; Jersey has slate stone roofs and terracotta tiles.

Figure 159 Conical roofs in Indonesia

Hungary and date palm branches in Israel. In India, we find the extreme ends from thin stone slabs used as roofing material in Himachal Pradesh, India. And the Taj Mahal is roofed with Marble, while on the other end, dried ruminant dung cakes in the villages. In Cameroon, they use banana leaves and even sun-dried mud for their huts. A turf roof (green roof) in Flam, Norway. In South Papua, the Kanume tribe uses *Melaleuca* sp tree bark, a.k.a. paperbark.

Figure 160 Selected global conical structures.

Figure 161 Dome roof structure was underway construction Ilog Maria Honeybee Farm

A dome roof structure construction was underway at the time of a visit to Ilog Maria Honeybee Farm km. 47 Aguinaldo Highway, Lalaan 1, Silang, Cavite, Philippines 4118, in Mar. 2020.

Figure 162 Conical roofs in other regions around the world.

Figure 163 Conical structures in Africa and other regions

Roundhouse and Conical concepts

Figure 164 Inspired by a Chiang Rai Bamboo village hut.

In keeping with the conical concept, illustrated herewith is a truncated conical structure of Chiang Rai, Thailand and the roundhouse of Nepal.

Goal Ghar of Nepal

Figure 165 Inspired by the Typical Nepali "Goal Ghar" or roundhouse

Figure 166 Log hive with a conical roof.

Novelty conical roof

The stingless bee log hives have a conical roof reminiscent of the *Mbaru Niang* of Flores Island. Heri Damora of Lampung, Sumatra did this style. He uses large dried leaves to add a cooling effect to the log hive. These are just novelties to complete the diverse collection.

Figure 167 Stingless bee keeping in South America

Beekeeping in the Neotropics

A contrasting method of meliponiculture is that the hives in logs are stacked horizontally on A-frames. It is an alternative way of harvesting the honey without upsetting the brood of the nest. Stingless beekeeping in South America takes shelter in conical huts. The ends are capped with a removable lid, the honey pots are pricked, and honey is poured out from the ends. The brood stays intact in the middle. The A-frame allows more log hives to be placed than standing vertically on shelves.

Conical huts of Papua.

1. *Honai* is the Papua traditional house that is home to the **Dani** tribe. *Honai comes from the words "hun" or male, and "ai" means home. Adult men usually inhabit Honai. Usually, Honai is found in valleys and mountains. The walls of this house are made of wood with a cone-shaped thatched* roof, at a glance, looking like mushrooms. The shape of this roof protects the wall's surface from rainwater and reduces the cold from the surrounding environment.

Figure 168 Honai Hut of the Dani Tribe in Papua

2. *Ebai* comes from the words "*ebe*", the body and "*ai*", which means home. Implying that this place is for women to live for life. *Ebai* is used to carry out the education process for girls, i.e., mothers will teach them things to do when they get married. *Ebai* is also a place of residence for mothers, daughters, and sons. But boys

Figure 169 Ebai hut for women of the Dani Tribe

who have grown up will move to *Honai*. The *Ebai* house is like the *Honai* but has a shorter and smaller size. Located on the right or left side of the *Honai*, the door is not parallel to the main door.

3. *Wamai* is a place that is used as a pet cage. Animals commonly used as livestock by tribes in Papua, such as chickens, pigs, dogs, and others. *Wamai* forms are usually square, but there are also other shapes,

Figure 170 Waimai hut for the pets, poultry and animals

very flexible depending on the size and number of animals each family owns.

4. *Kariwari* is a Papua traditional house inhabited by the **Tobati-Enggros** tribe, who live on the shores of Lake Sentani, Jayapura. This house is a special home for men who are around 12 years old. This house is used to educate these children - Lessons learned include building boats, how to fight, making weapons, and

Figure 171 Kariwari for children and communal events.

sculpting. This house has an octagonal shape that resembles a pyramid. This form was made to be able to withstand strong gusts of wind. While the conical roof, according to the people's belief, gets closer to their ancestors. The height of this house varies from 20 to 30 meters. Consists of 3 floors that have their respective functions. The bottom floor is used as a place to study boys.

The second floor is used as a meeting room for leaders and tribal chiefs and as a place to sleep for men. The third floor is a place for meditation and prayer. Its function is that the roof is not separated and flown by the wind. Under the log, flooring is used to store handicrafts, tools of war and others[44].

3- tiered conical roof series

This conical three-tiered green roof structure is designed for a Bee Sanctuary or Repository to accommodate beehive placement at three levels. It allows for bee forage within proximity while keeping safe from flood levels and wild animals.

Figure 174 Inspired by the Honai, a Papua traditional house that is home to the Dani tribe.

The Green Roof Bee House (Figure 145) is on a raised platform elevated above prevailing flood levels. It also served as a barrier to wild animals and was inspired by a carnival carousel. The Round Barn (Figure 145) on a raised concrete curbed foundation to safeguard the basement from flooding and also a vermin barrier

Figure 172 3- tiered conical roof series

---

[44] Web reference https://cuorerosanero.com/5-papua-traditional-house/ Indonesian culture - 5 Papua traditional house July 23, 2020, Mulyadi Azis

Chapter 14

Hanging hives and host containers
By Abu Hassan Jalil

Dried Gourds, Winter melons, Squash, Pumpkins, and the globular *Maja* fruit or Indian Bael (*Aegle marmelos* (L.) Correa) as in Lombok Is. and NTB region of Indonesia.

Figure 173 Dried Gourds, Melons, Squash, Pumpkins and Maja

Coconut Kernel single or stacked as in Silay City in Negros Occidental & Dumaguete in Negros Oriental in Visayas Region of the Philippines. In Malaysia, fallen coconuts hollowed by squirrels are used by SB feral hives. These are hung around villagers' house roof eaves.

Figure 174 Coconut kernels

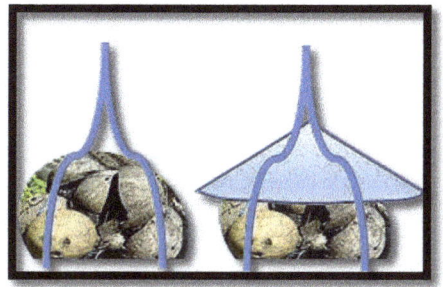

Figure 175 Coconut shell halves - Bao Technique

Coconut shell halves – *Bao* technique of Bicol region of the Philippines. Maybe exposed or covered with a metal sheet cone or bundled with *anahaw* fan-palm leaves. This hiving technique is common in the Bicol region and sometimes adopted in Northern Samar, Mindoro, and Marinduque Islands.

Bamboo hives individually hung or multiple hives of Borneo and Vietnam. These are predominant in areas where feral colonies are found in bamboo groves.

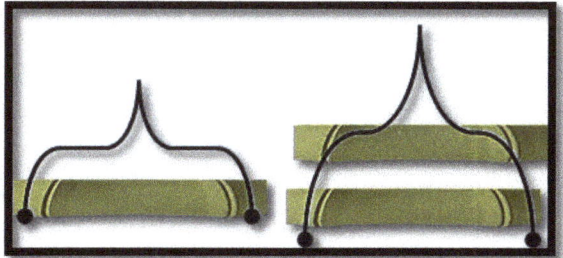

Figure 176 Bamboo Hives

Branch hollows individually hung or multiple hives as in Kerala, S. India and Bali. In Malaysia, we used to get feral colonies in branch hollows in some graveyard vicinity or compound.

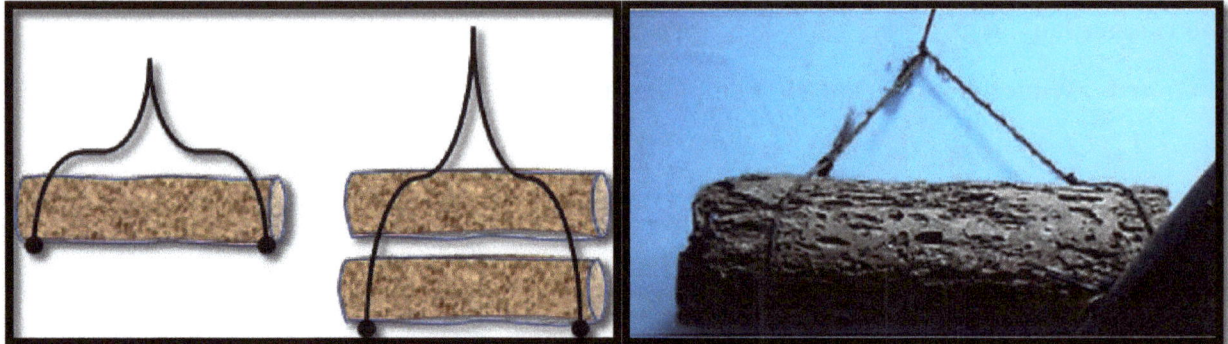

Figure 177 Branch hollow Beehives

Wharf or jetty floats (buoys from Polillo Island). We saw these in PJ plantation, Anos, Quezon Province, Philippines. The beekeepers cut a flap at the upper back to insert brood cells and drill a hole at the front for the nest entrance.

Figure 178 Wharf or Jetty floats beehive.

PVC pipes of varying diameters emulate tree cavities that the bees favour. These hives, as seen in Kerala, South India, were innovated by Binu PT,

Figure 179 PVC Pipe Hive.

Clay pots or **terracotta** pots can be hung on roof eaves. Clay pots are used in Gn. Kidul & Pati in Central Java; Indonesia; Sabah, Malaysia; Pondicherry, India and Sagada & Bohol Philippines.

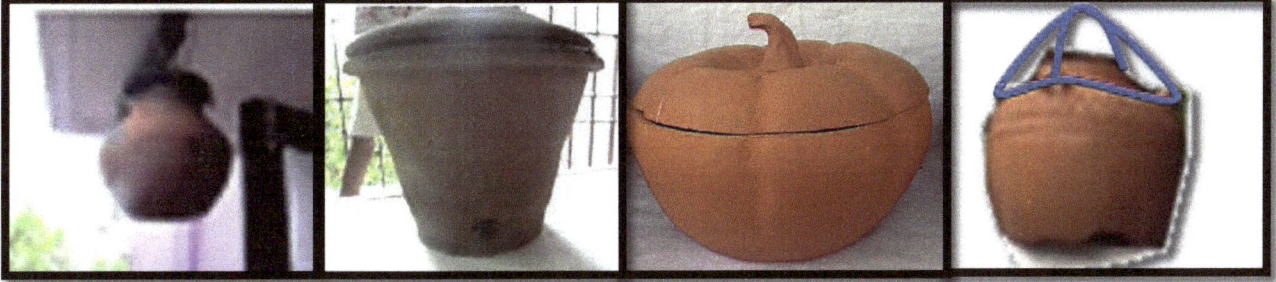

Figure 180 Typical Sagada clay pot with ears for hanging

Figure 181 Log Hive

Log hives are being hung to avoid frogs, ants, lizards, and other crawling enemies. Mostly seen in parts of Borneo and Malaysia.

Hanging box hives as being practised in Many areas in the Indo-Malaya Eco-Zone.

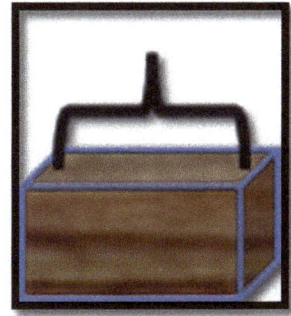

Figure 182 The wooden Box hive

Hanging coffins

In the Philippines, hanging coffins are one of the cultures, namely among the Sagada people, who have been cultivating through the years with the belief that their loved ones will be closer to their ancestral spirits. However, they stopped it last 2010. They now prefer having their deceased loved ones buried where they can visit and reminisce memories with them. Besides being on the topic of hanging containers, there are two other reasons why this item is in this book. One is that it would be a target for necrophagous bees, and secondly, it indirectly relates to ancestor veneration.

Figure 183 Hanging coffins are one of the cultures of Sagada, Mountain Province, Luzon.

In Indonesia, hanging coffins (*Liang tokek*, literally "hanging burial") is one of the funerary practices of the **Toraja** people at the Londa Nanggala Cave in Sulawesi, either for primary or secondary burials. The reasoning for their placement is to discourage looters who might steal the items interred with the dead. The distinctively boat-shaped coffins, *erong*, are always placed below overhanging parts of the cliff face. These can be natural overhangs or cave openings, but some coffins are placed beneath

Figure 184 Funerary practices of the Toraja people at the Londa Nanggala Cave in Sulawesi

man-made overhangs. They are guarded by carved wooden representations of the dead, known as *tau-tau*. Older *tau-tau* is more abstract, but more modern *tau-tau* can be lifelike.

Liang tokek was reserved for the "founders" of the village and thus is among the oldest dated coffins, dating to around 780 AD. Like the hanging coffins of the Philippines, Liang tokek accounts for only a minority of the region's funerary practices. They were part of burial complexes, which included other interment practices, usually differing based on the social class and age of the dead. The Torajans believe these complexes to be abodes of spirits of the dead in the afterlife.

Figure 185 Famous hanging coffins are those which were made by the Bo people (now extinct) of Sichuan and Yunnan

In China, hanging coffins are known in Mandarin as *xuanguan* (simplified Chinese: 悬棺; traditional Chinese: 懸棺; pinyin: *xuán guān*), which also means "hanging coffin". They are an ancient funeral custom of some ethnic minorities. The Bo people (now extinct) of Sichuan and Yunnan made the most famous hanging coffins. Coffins of various shapes were mostly carved from one whole piece of wood. Hanging coffins lie on beams projecting outward from vertical cliff faces, placed in cliff caves or on natural rock projections on steep mountainsides. Another hanging tomb (an example of an exposed natural site) is of the **Ku** People at Bainitang (白泥塘), Qiubei County, Wenshan Prefecture, Yunnan Province, China.

Mobile Racks for Beehive Housing

Figure 186 Moveable flower cart that incorporates a Bee shack on a Bullock cart.

Figure 187 Inspired by a Mongolian Yurt on a Nomadic Trailer

The concept of moveable bee shelters is ideal in situations of impending floods. They may be carted by beasts of burden or motorised equipment. They also will serve well for beekeepers practising pollination services where the hives must transport to different farm locations. They can be in the form of caravans or gipsy *vardo*[45], roadshow buggies on trailers, campers, nomadic tents or yurts, bullock carts or horse carriages.

Figure 188 Moveable Bee House on a trailer – Inspired by a Gypsy Caravan.

Figure 189 Artist's Bee Shed on a trailer – Inspired by a roadshow buggy.

This road show buggy model is excellent for pollination. It can be parked at any location and later moved along any paved road. Potted flowering plants can be carried along for a time during transit.

---

[45] a vehicle like a caravan used by Romani people; a Gypsy caravan.

Chapter 15

The advent of Green Roofs

Part 1 of Volume 1, 'Conservation in Meliponiculture', encourages the closeness of foraging sources. One cannot get any closer than at the foot or the roof of your bee rack or shack. A roof is a top covering of a building or structure, including all materials and constructions necessary to support it on the walls of the building or uprights, protecting against rain, snow, sunlight, extremes of temperature, and wind. A roof is part of the building envelope.

The characteristics of a roof depend upon the purpose of the building that it covers, the available roofing materials and local traditions of construction. Along with architectural design and practice concepts, they may also be governed by local or national legislation. In most countries, a roof protects primarily against rain. A veranda may be roofed with material that protects against sunlight but admits the other elements. The roof of a garden conservatory protects plants from cold, wind, and rain but admits light. A roof may also provide additional living space (for the bees), such as a roof garden.

Figure 190 Inspired by the original bonnet roof and split into three levels for the bees to forage independently on each level.

Commercially available Drainage boards with geotextile lining and lightweight soil mixes (peat fibre with perlite, vermiculite or pumice) are an absolute necessity for modern roofs.

In the old days, planting on rooftops was already practised but with conventional methods incorporated into a slope or inclined embankment.

Figure 191 Initially inspired by the Mabuhay house in Samar Island of the Philippines and seems to have similarities to Viking Norse houses.

Such green roof systems applied for Bee sheds in Meliponiculture have great advantages:

Advantages

1. Consolidation of safety and comfort factors.
2. Transition bay before exposure to the general population on a bee farm
3. Hospitals for ailing colonies
4. Quarantine of contaminated hives
5. Centralised harvesting
6. The higher benefit of foraging space cost
7. Foraging in all weather conditions (Typhoons not included)
8. Better overall thermal comfort for bee colonies

Disadvantages

1. Not feasible for farm produce pollination program
2. Construction Costs and precise technical application
3. Load-bearing strut bracing and trusses are required for water-soaked soil on the roof

Figure 192 The 'Hobbits' and some Vikings had a similar cool idea.

The modern turf roof in Norway is a testament to the Nordic Viking heritage. Some new buildings even incorporate solar panels beside the turf.

Figure 193 A Modern Turf roof in Norway, a testament to the Nordic Viking heritage.

Beengo Farm is a Bee and Mango Farm located in the smallest town by area and population in Tunga, Leyte, in Eastern Visayas. Owned and managed by Gary A. Ayuste, better known as "Mr Garybee".

Figure 194 The 'Hobbit' house in Beengo Farm in Leyte, Eastern Visayas.

Gary Avila Ayuste is one of the more successful Stingless Bee keepers in Leyte Island, Philippines. Gary captured the vernacular ambience elegantly and built a "Hobbit" house with a green roof. The house is partially sunken, so maintenance of the plantings on the roof (at average human chest height) is much easier. His place attracts quite a stream of local and foreign tourists.

Figure 195 Gary built a traditional bee house by the creek that runs by his farm.

Incidentally, the traditional Mabuhay house of Samar Island, the Philippines, is ideal for a green roof, as illustrated below.

Figure 196 Mabuhay house on Samar Island and a Green roof bee shack.

The roof eaves need not reach the ground height that can be made up by stacking gunny sacks filled with soil to grow flowering creepers and climbers that can creep up the roof.

Figure 197 Inspired by the traditional Mabuhay house of Samar Is.

Green Roofed Boathouse[46]

Functional to the environment and human needs and stripped of the idea of a fixed land, Houseboats can produce their water, recycle their effluents and waste, and offer a green roof where it is possible to grow different options and serve as self-sustaining Beekeeping facilities.

In turn, if the predictions about global warming are true, it is possible that sea levels will rise and that currently inhabited cities will be invaded by water.

---

[46] Source: www.labioguia.com, http://www.econautico.com.ar/one

Flood-protected hospitals, greenhouses, and even research repositories are some of the projects that may use a floating platform constructed on water.

Although many places use this type of house, forming floating neighbourhoods, and its existence goes far in time, this proposal can be sustainable and tempting for souls seeking innovative ways of life. It is adaptable to various places, and it is even possible to find various ways of building them on the web, which promise to be accessible to modern man.

Most of the projects base their forms on the use of solar panels that include a tidal energy harnessing system that supplies all the needs of the house, both day and night.

Figure 198 Green Roofed Boathouse model examples (excerpts from https://medium.com/goplaceit-datos/fant%C3%A1sticas-casas-flotantes-e9d697caaac8

Chapter 16

Islamic Structures and Mosques
By Abu Hassan Jalil

Working with straight beams for fast and easy construction allowed us to explore contemporary mosque designs. One may choose a model of the King Faisal Mosque in Islamabad, Pakistan, to incorporate the Bee Box Hive Roof Structure and direct sun protection. This model is something for Muslim Beekeepers anywhere in Indo-Malaya.

Figure 200 Inspired by the King Faisal Mosque in Islamabad, Pakistan.

The roof design is a four-sided Butterfly roof. The finial on the roof summit and each minaret is an upturned crescent. This crescent makes for a perfect hook for the shade over the whole model. The roof can be wooden and covered with linoleum. The pediments at the gable ends can be perforated with acrylic or Perspex sheets. The shade can be 80% black netting over a plastic sheet to protect from rain and direct sun.

Figure 199 Some of the Mosques in Thailand. Images from Wikimedia

There are many Mosques in Thailand (Figure 172), and we find Islamic calligraphy and Islamic design adornments.

Figure 201 On the left are Calligraphy and Islamic designs in Thailand. A mosque-inspired bee housing on the right in north Peninsula Malaysia.

On the left are Calligraphy and Islamic designs in Thailand. A mosque-inspired bee housing on the right in North Peninsula Malaysia (Figure 174).

Inspiration from religious structures.

Figure 203 Left: Tranquerah Mosque in Malacca; Right: Proposed tiered three roof bee house/

Figure 202 Pyramidal top mosques in Melaka and Riau.

Traditional Vernacular Style – A three-tiered pyramidal roof like the Masjid Trengkera (on Tranquerah Street, Melaka) is believed to have been built as early as the 16th century. We see many more examples throughout the Nusantara region as late as the 18th century, including on Sulu Island, Philippines. Indeed, the top of the Nusantara traditional mosques in the 18th century is not in the shape of an onion dome but the shape of a 4-sided pyramid. The same goes for the top of the minarets. However, in Thailand, mosques are in the shape of onion domes (see page 31) ... this may be so because, in ancient times, Islam's influence was not as widespread as in the 18th century.

Figure 204 Sheikh Karimul Makhdum Mosque. Tawi Tawi Island

Figure 205 More examples of three-tiered roofing of mosques in Sulawesi and Lampung

Figure 207 Green roof bee house inspired by pyramidal tired roof mosques.

I remember my younger days; I have always been intrigued by the history of civilisation in Mesopotamia. Babylonia was a state in ancient Mesopotamia. The city of Babylon, whose ruins are located in present-day Iraq, was founded more than 4,000 years ago as a small port town on the Euphrates River. Babylon, located about 80 km (50 miles) south of modern Baghdad in Iraq, was an ancient city with a settlement history dating back to the 3rd millennium BCE.

Figure 206 Ziggurat Bee House, a multi-tiered Green Roof structure inspired by "The Hanging Gardens of Babylon".

Hanging Gardens of Babylon' was probably the 19th century after the first excavations in the Assyrian capitals. The Hanging Gardens of Babylon in this hand-coloured engraving probably was made in the 19th century after the first excavations in the Assyrian capitals. It depicts the fabled Hanging Gardens of Babylon, one of the World's Seven Wonders. According to tradition, the gardens did not hang but grew on the roofs and terraces of the royal palace in Babylon. Nebuchadnezzar II, the Chaldean king, is supposed to have had the gardens built in about 600 BC as a consolation to his Median wife, who missed the natural surroundings of her homeland.'' (Unknown Author in Public Domain)

Figure 208 This hand-coloured engraving, probably made in the 19th century after the first excavations in the Assyrian capitals, depicts the fabled Hanging Gardens, with the Tower of Babel in the background. Public Domain File: Hanging Gardens of Babylon. Jpg Created: 19th-century; Unknown author - http://www.plinia.net/wonders/gardens/hgpix1.html

That was my initial motivation for beginning a career in Landscaping. Gardening was a passion I had as a kid till today. This inspiration drove me to design the Ziggurat multi-tiered green roof Beehouse (Figure 179). I included water tanks at the top and the front porch awning canopy. At the apex is a clear Perspex pyramid for a night beacon. Each storey top will have planter pots filled with lightweight soil, and the whole system will be connected with a drip irrigation reticulation. The base is adjoined to ground beds of creepers, climbers and flowering ramblers. These vines can creep up to the final tiered inclined roof.

With sufficient balconies and well-insulated sloping roofs, the bee hives can be easily placed or hung for convenient foraging sources. In keeping with Babylonian traditions, the green roofs can be mulched with Date palm (*Phoenix* sp.), spent fruit bunch and stalks or mashed frond petiole keels practised in Israel as thatch roofing.

Traditional houses with balconies and a veranda make for ideal Bee Galleries. Sufficiently open for maximum airflow and elaborate traditional shaped roofing, allowing hot air dissipation. This example (Figure 182) is earmarked for a partial green roof at the veranda foyer.

Figure 209 Traditional house with balconies and a veranda in Negeri Sembilan, Malaysia

Chapter 17

Beescape and Wind Effects

Beescape is landscape planning that facilitates bee forage ahead of aesthetic values. As wind velocity is essential in meliponiculture, effective and strategic placement of windbreakers is paramount in providing resource portioning in bee foraging.

Eddies and Scales of Turbulence. Turbulent flows develop spinning or swirling fluid structures called eddies. Eddies can stretch, merge, and divide; such eddying motion is a characteristic feature of turbulence. Turbulent flow is, therefore, rotational and has vorticity.

Figure 210 Wind Breakers and Effects

Friction between the moving air mass and the earth's surface features (hills, mountains, valleys, trees, buildings, etc.) is responsible for the swirling air vortices, commonly called eddies. They vary considerably in size and intensity depending on the surface obstruction's size and roughness, the wind's speed and the air's stability. They can spin in either a horizontal or vertical plane. Unstable air and strong winds produce more vigorous eddies. In stable air, eddies tend to dissipate quickly. Eddies produced in mountainous areas are especially powerful.

GUSTINESS

A gust is a rapid and irregular fluctuation of varying intensity in air currents' upward and downward movement. It may be associated with a rapid change in wind direction. Gusts are caused by mechanical turbulence that results from friction between the air and the ground and by the unequal heating of the earth's surface, particularly on hot summer afternoons.

SQUALLS

A squall is a sudden increase in the strength of the wind of a longer duration than a gust and may be caused by the passage of a fast-moving cold front or thunderstorm. Like a gust, it may be accompanied by a rapid change of wind direction.

DIURNAL VARIATIONS

Diurnal (daily) variation of wind is caused by strong surface heating during the day, which causes turbulence in the lower levels. The result of this turbulence is that the direction and speed of the wind at the higher levels (e.g., 3000 feet) tend to be transferred to the surface. Since the wind direction at the higher level is parallel to the isobars, and its speed is greater than the surface wind, this transfer causes the surface wind to veer and increase speed.

GALE

A gale is a strong wind, typically used as a descriptor in nautical contexts. Gale is a wind stronger than a breeze; specifically, a wind of 28–55 knots (50–102 km per hour) corresponding to force numbers 7 to 10 on the Beaufort scale.

The word gale possibly originates from the Old Norse word galinn[47], which means "mad", "frantic," or "bewitched." Weather forecasters sometimes use the term "gale-force winds" to describe conditions that aren't quite as extreme as hurricanes or tropical storms but probably fierce enough to snap your kite in half.

PREVAILING WINDS

A wind from the predominant or most usual direction at a particular place or season. For example, "the trees all leaned inland, away from the prevailing wind."

---

[47] Galinn (died 873) was a Viking who served under Soma in Anglo-Saxon England during the 9th century. Alongside Birna and Lif, he was one of Soma's closest associates. Ultimately, he betrayed Soma in order to pursue his own "destiny".

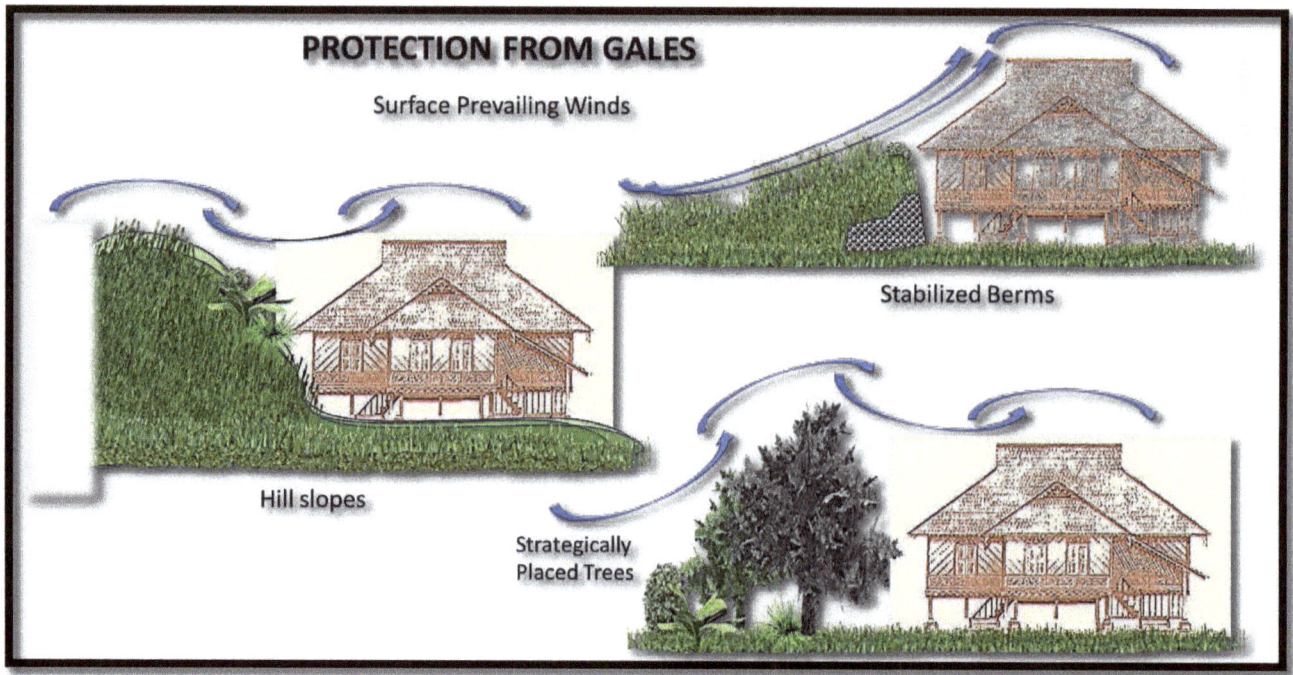

Figure 211 PROTECTION FROM GALES

Figure 212 PROTECTION FROM STRONG WINDS

Protection From Gales

We may adopt some protection methods if we know the surface prevailing winds.

1. Adjacent Hillock – building adjacent to small hills or hillocks to shade in the opposing direction.
2. Stabilized Berms – With no hills available, one may raise berms or bunds securely stabilized with rubble walls or gabions.
3. Strategically Placed Trees – Incorporating Beescape appropriately may divert strong winds while adding forage sources for the bees.
4. Thick Walls – A good example is the Ivatan House, Batanes Island. These limestone-walled houses have been known to withstand the worst of typhoons.
5. Boulder Soil Mound – Rock outcrops raised with soil mounds and boulders; an example is a Moro Mosque in Mindanao.
6. Slanting Roof – Like the Mabuhay House of Samar Island, it can deflect strong winds and keep the main structure safe.

Beescape Improves Internal Air Circulation

This air circulation is accomplished by consideration of the building's overall shape and the positioning of windows and doors. Proper placement of trees and shrubs can and does provide for deflection and redirection of wind. The diagrams below show how ventilation is achieved using natural elements.

We better understand wind and air movement by experimenting with Beescape for Meliponaries and Bee galleries. Realizing that space is also an element in landscaping, one can visualize airflow. By manipulating the placement of plants and bee forage provisions, one can direct, divert and redirect the airflow necessary to achieve thermal comfort.

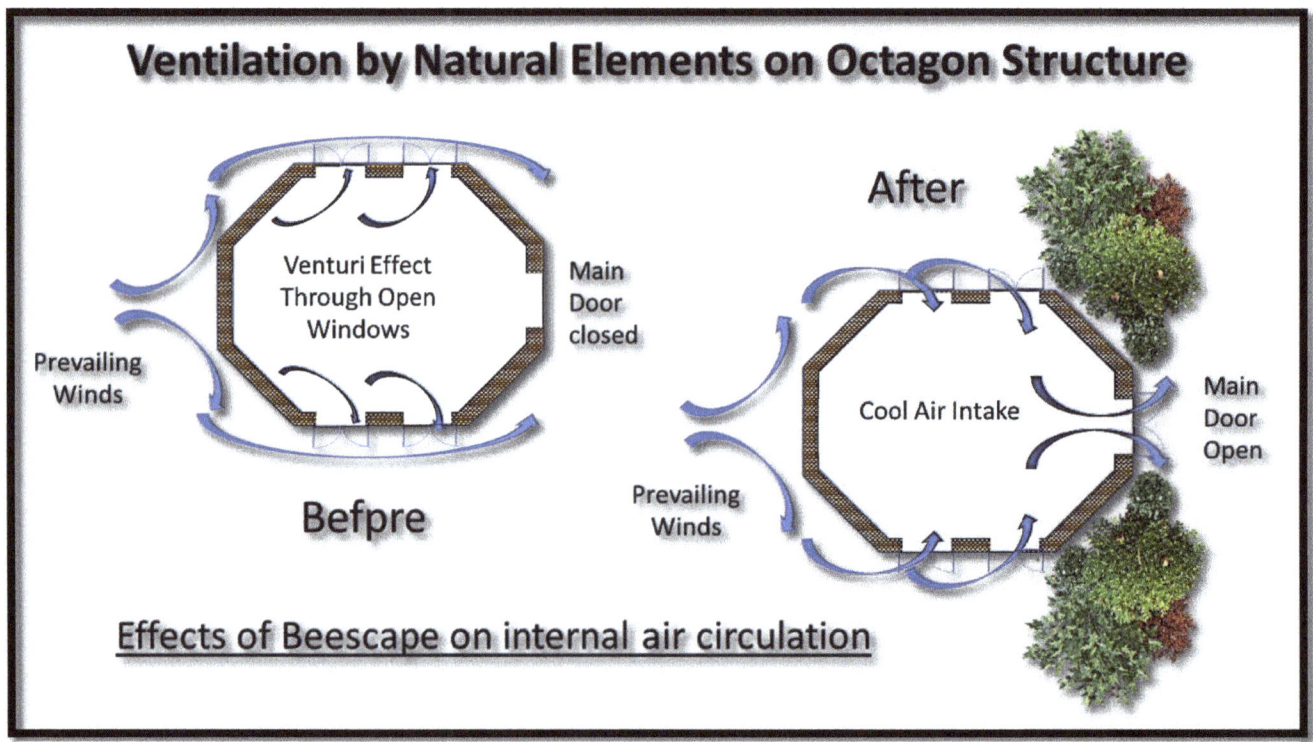

Figure 213 Ventilation by Natural Elements on Octagon Structure - Effects of Beescape on internal air circulation

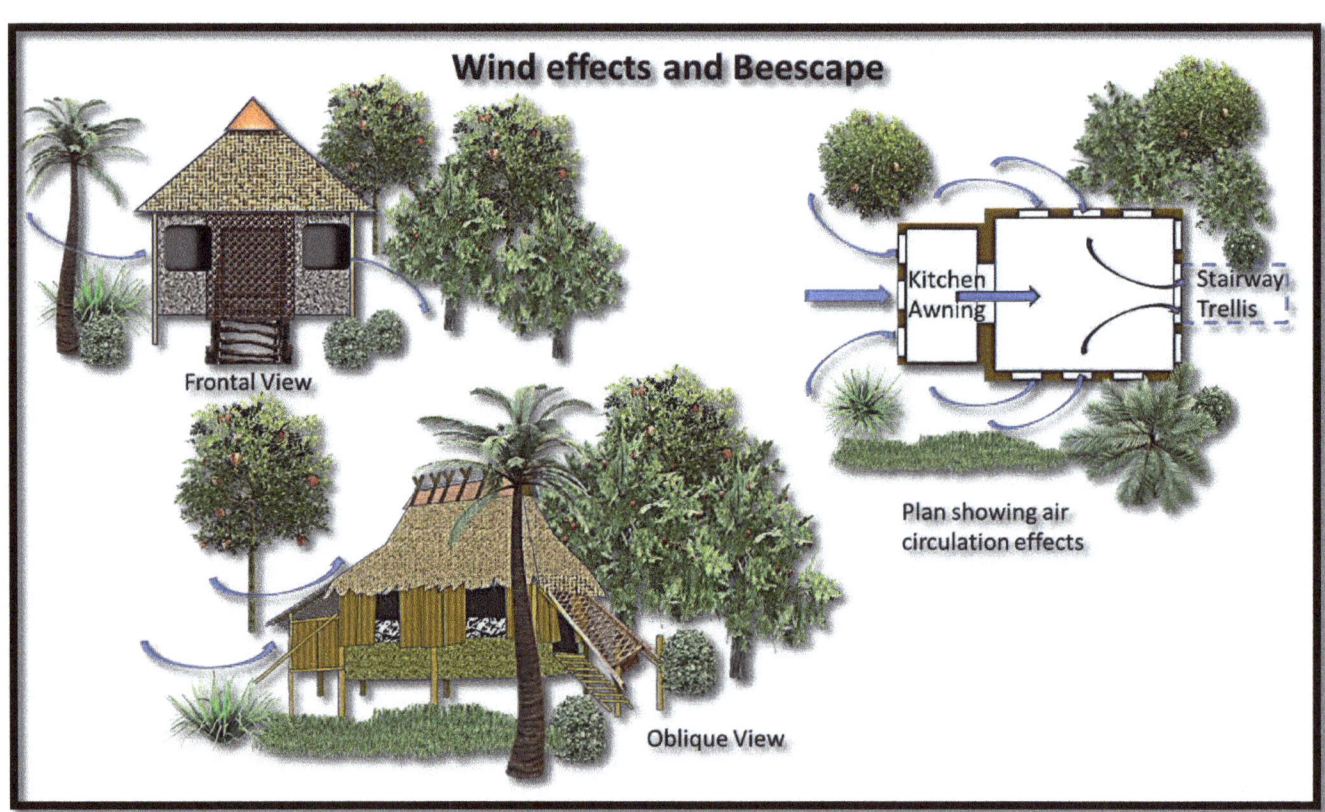

Figure 214 Air circulation and Beescape for Ibaloi House

List of Figures

Figure 1 Vernacular architecture in the Philippines. ... - 11 -
Figure 2 Almost completely enclosed roofing structures for protection against extreme weather. ... - 12 -
Figure 3 Example Vernacular in Global Meliponaries. ... - 12 -
Figure 4 Aguaruna people building their houses in Bajo Cachiaco, Peru. ... - 13 -
Figure 5 Typical Tausug Architecture influence on this ancestral home. ... - 13 -
Figure 6 The uniquely constructed house in Palawan ... - 14 -
Figure 7 Leo Grajo's new setup devastated after Typhoon ... - 14 -
Figure 8 Traditional Huts and Houses I photographed during travels in The Philippines. ... - 15 -
Figure 9 Hexagon Beehouse in Grajo's Bee Farm in Sorsogon. ... - 16 -
Figure 10 Modifications on the Hexagon Bee house in Grajo's bee farm, Sorsogon ... - 16 -
Figure 11 Gazebos, Cabanas and Pavilions as Bee sheds or Bee rack shelters. ... - 17 -
Figure 12 Lakeside Two-Tiered Bee Pavilion ... - 18 -
Figure 13 Balinese architecture inspired Bee Rack ... - 18 -
Figure 14 Danny Hizon's house in Sabluyan, Occ. Mindoro; Right: A Balinese structure that inspired Danny Hizon ... - 18 -
Figure 15 The Ivatan House a.k.a. Kalinga House in Batanes Island, Philippines. A Sinadumparan Ivatan house, one of the oldest structures in the Batanes Islands. The house is made of limestone and coral and its roofing is of cogon grass. ... - 19 -
Figure 17 Inspired by the Ivatan limestone rubble wall house. ... - 20 -
Figure 16 Ifugao house or Igorot Hut ... - 20 -
Figure 18 The Kankanay House ... - 21 -
Figure 19 Ibaloi house. ... - 21 -
Figure 20 The Isneg house. ... - 21 -
Figure 21 Gilitob - Octagonal House of the Kalinga. ... - 22 -
Figure 22 The Tinguian House ... - 23 -
Figure 23 The traditional Bontoc house ... - 23 -
Figure 24 The Aeta's house ... - 24 -
Figure 25 The Mangyan House ... - 24 -
Figure 26 The Tagbanwa House ... - 24 -
Figure 27 Maranao Carvings. ... - 25 -
Figure 28 Left: Sarimanok or (Papanoka Mra) is a legendary bird of the Maranao; Middle: A Maranao house preserved amidst modern buildings in the background at Marawi City, Lanao del Sur, Mindanao, Philippines; Right: Model of Torogan house of the Maranao people at Cockington Green Gardens ... - 25 -
Figure 29 A 300 yr old Maranao House. Inset: Top right: Human scale for the size of the tree trunk post; Bottom right: closeup of stone castors beneath the Tree trunk pillars. ... - 25 -
Figure 30 Inspired by the Traditional Maranao house with rock castors beneath tree trunk posts. ... - 26 -
Figure 31 The Classic Maranao Torogan ... - 26 -
Figure 32 Maranao Royal House with Princess' Chamber ... - 26 -
Figure 33 Maranao Bee Gallery ... - 26 -
Figure 34 Modified Maranao Torogan ... - 26 -
Figure 35 The Yakan house ... - 27 -
Figure 36 A Tausug House with 109 tiny SB colonies in the bamboo walls. ... - 27 -
Figure 37 The Bajao House ... - 27 -
Figure 38 The Mandaya House ... - 27 -
Figure 39 The Gono Taug'na House of the T'boli Tribe ... - 28 -
Figure 40 Tog'gan House of the Mangguangan Tribe ... - 29 -

Figure 41 Uyaanan House of the Mansaka Tribe .... - 29 -
Figure 42 Left: Gantangan is an old measurement of selling rice in the absence of a weighing scale. Contributed by Alberto V. Bartolata; Right: . Ata Paquibato - Gantangan Palabain ref: https://m.facebook.com/watch/?v=1363139273747172&_rdr .... - 30 -
Figure 43 Binotok House of the Ata Manobo Tribe... - 30 -
Figure 44 Bakag Farmhouse of the Obo Manobo Tribe .... - 32 -
Figure 45 The traditional house of the Bagobo Tagabawa of Tibolo, Sta. Cruz. .... - 32 -
Figure 46 The Gumne House of the B'Laan Tribe .... - 33 -
Figure 47 The Bal'Lay House of the Mandaya Tribe .... - 34 -
Figure 48 Pre-colonial Moro Mosque at Lake Lanao, Mindanao .... - 34 -
Figure 49 Bahay Kubo in the Maharlika series .... - 35 -
Figure 50 Maharlika Bee House on Hill Slope .... - 36 -
Figure 51 Inspired from a Maharlika Datu House Painting by JBulaong 2019 .... - 36 -
Figure 52 Inspired by a Pre-colonial Maharlika Mountain House painting by JBulaong 2020 .... - 36 -
Figure 53 Inspired by a JBulaong 2020 painting of a Pre-colonial Maharlika Split Level Upland House .... - 36 -
Figure 54 Inspired by a Pre-colonial Maharlika Lowland House painting- by JBulaong 2020 .... - 36 -
Figure 55 Inspired by a Pre-colonial Maharlika Lakeside House painting by JBulaong 2020 at Lake Sebu, Mindanao .... - 36 -
Figure 56 Maharlika Grain store Adopted Bee House .... - 37 -
Figure 57 Bee Shacks done by Agriculture Coops in Marinduque Island .... - 37 -
Figure 58 A small Subanon village hut on Mount Malindang .... - 38 -
Figure 59 Ancestral domains of the Subanon in Zamboanga peninsula and northern Mindanao .... - 38 -
Figure 60 Political map of the Zamboanga Peninsula, Philippines. Showing Zamboanga del Norte, Zamboanga del Sur, Zamboanga Sibugay, Isabela City, and Zamboanga City. .... - 39 -
Figure 61 Tetragonula dapitanensis queen and worker illustration .... - 39 -
Figure 62 A typical Tagbanwa hut .... - 40 -
Figure 63 Palawan Stingless Bees cited in Cockerell 1919 .... - 40 -
Figure 64 The ancestral home of Lucia Gadian of Brgy. Awis, Leon, Panay Island .... - 41 -
Figure 65 The 60 yr. old bamboo house of Antonio Capirayan of Brgy Tacuyong Sur, Leon, Iloilo, Panay – almost collapsing. .... - 41 -
Figure 66 Bamboo house of Esperidion Calaor ng Brgy. Gumboc, Leon, Iloilo, Panay also houses 24 bee colonies .... - 42 -
Figure 67 A bee hut in Romblon .... - 42 -
Figure 68 A rest Hut at TESDA RTC garden, Iloilo .... - 43 -
Figure 69 Nora Peña's Bee Sanctuary and a bee rack in Leon, Iloilo .... - 43 -
Figure 70 Sagada House in Tublay, Benguet, Central Luzon. .... - 44 -
Figure 71 Bee housing by Cebu Stingless Bees .... - 44 -
Figure 72 images are a Bahay Kubo on the left, an ornamental bamboo hut at Nisol Farm, Gawahon and an Okinawa house giving a Japanese accent in the Eco Park .... - 45 -
Figure 73 Rest hut in Balai Buhay sa Uma .... - 45 -
Figure 74 Novelty bee housing in Los Pepes, Cavite; Marl's Bee store housing his feral colonies .... - 45 -
Figure 75 Left and right images are restrooms and the big one in the middle is a Bee gallery. .... - 46 -
Figure 76 Rest huts made of bamboo and we found several nest entrances built in the joints and the bamboo structures... - 46 -
Figure 77 Philippines Honeybees Industries (PHI) Bibera Research Training Centre in Pilar, Sorsogon, Bicol Region, ... - 47 -
Figure 78 Hive racks and a Hexagon Bee House .... - 47 -
Figure 79 The Bee Gallery at Pia's Bee Farm and the two-level structure by night. .... - 48 -

Figure 80 Owners are Peter Paul Rafasola and his wife Claire Azuelo-Rafasola are former beekeepers in America. The box hive construction appears to have an Australian design influence with a clear plastic cover for observation. Encountered Tetragonula hive in a woven bamboo wall. .... - 49 -
Figure 81 Mikee was badly hit by the Typhoon, as shown by his Aunty Feli .... - 50 -
Figure 82 Mikee Balais' Meliponary in traditional architecture with a 'green roof' for his shack. .... - 50 -
Figure 83 Example of an octagon house of the Kalinga tribe in Luzon .... - 51 -
Figure 84 Inspired by the Bongao Municipal Hall, Tawi-Tawi .... - 51 -
Figure 85 Subregions of Oceania. .... - 56 -
Figure 86 Native houses in Aru Islands .... - 57 -
Figure 87 Left: Hoise emblem of Koror State Flag; Right: Island of Yap, Federated States of Micronesia, Caroline Islands, Pacific .... - 57 -
Figure 88 Top Left: A traditional Palauan bai: Top Right: Village on the Palau Islands, painting by Rudolf Hellgrewe c. 1908.; Bottom left: Palau International Airport; Bottom Right: Flag of Koror State Palau .... - 58 -
Figure 89 Old Australian cottage design housing Tetragonula hockingsi in Brisbane belonging to Bob Luttrell. - 59 -
Figure 90 I recorded this modern version of a cottage. And this post-modernist (Clark Kent's) Telephone booth. .... - 59 -
Figure 91 Some classical designs of Australian Bee housing. .... - 60 -
Figure 92 Log Cabin–Shaped Logs .... - 60 -
Figure 93 Austroplebeia australis (Friese, 1898) .... - 61 -
Figure 94 Austroplebeia cassiae (Cockerell, 1910) .... - 61 -
Figure 95 Austroplebeia essingtoni (Cockerell, 1905) .... - 61 -
Figure 96 Tetragonula clypearis (Friese, 1909) .... - 62 -
Figure 97 Tetragonula mellipes (Friese, 1898) .... - 63 -
Figure 98 Austroplebeia magna Dollin, Dollin & Rasmussen 2015 .... - 63 -
Figure 99 Distribution of Australasian and Papuasian Stingless Bees in Wallacea & Sahul .... - 64 -
Figure 100 Tetragonula sapiens (Cockerell 1911) .... - 65 -
Figure 101 *From Left, House on stilts at the lagoon of Abaiang atoll; A house of Tebunginako village; Kiribati style church in South Abaiang; 2 storey thatched house in South Abaiang* .... - 65 -
Figure 102 Medium shot of idyllic village beach near Auki, the capital of Malaita. .... - 66 -
Figure 103 Recent Earthquakes on the Solomon Islands as of 22, Nov. 2022 .... - 66 -
Figure 104 a Vale Vakaviti and a Fijian Bure .... - 67 -
Figure 105 Levuka, 1842... The first Europeans to land and live among the Fijians were shipwrecked sailors like Charles Savage. .... - 67 -
Figure 106 Traditional house construction on the island of Aneityum. Photograph: Gregory Plunkett ...... - 68 -
Figure 107 Left: Historical image from 1875 showing one of the earliest representations of the nakamal in Vanuatu; Middle: The Taloa Farea on Nguna Island post Tropical Cyclone Pam; Right: The Moriu Nakamal on Epi Island post Tropical Cyclone Pam .... - 68 -
Figure 108 Left: Marketplace of Kalo, British New Guinea, 1885; Right: Lae International Hotel, 4th Street Lae Archway .... - 69 -
Figure 109 Tribe in Papua New Guinea with wigwam at a settlement; Source: 1889 Geographie .... - 69 -
Figure 110 Austroplebeia cincta (Mocsáry in Friese, 1898) .... - 70 -
Figure 111 The National Emblem of PNG .... - 71 -
Figure 112 Bevak (traditional home of Kanume tribe South Papua) done by Theresia Agnesia Maturbongs aka Hesty and Madu Pokos Merauke— at Bevak Lebah Tetragonula Cf Mellipes _ Simpang Yanggandur Rawa Biru, Yanggandur, Merauke, Papua. .... - 72 -
Figure 113 Manokwari House of West Papua .... - 73 -

Figure 114 Asmat on the Lorentz River, photographed during the third South New Guinea expedition in 1912–13. ..... - 73 -
Figure 115 Traditional; River mangrove or estuary village of the Asmat people of Papua, Indonesia, ... - 74 -
Figure 116 Asmat shields ..... - 74 -
Figure 117 Wood carving depicting Asmat Hut and daily life ..... - 74 -
Figure 118 Vernacular dwelling structures in Benteng None Village, West Timor, NTT. ..... - 75 -
Figure 119 Left: Houses in Fatumnasi Highland region; Right Tribal house in Amarasi, Kupang Province ... - 76 -
Figure 120 Left: Fatumnasi Rumah Bulat (Roundhouse); Right: Tamkessi Conical Hut in West Timor. ..... - 76 -
Figure 121 A traditional house called laco, for meeting the families that are part of a Lulic (sacred) house and also to receive guests, circa 1968–1970. ..... - 77 -
Figure 122 Inspired by the Uma Lulik or Sacred House of Timor Leste. ..... - 77 -
Figure 123 Left: Holy House in Maununo, Suco Cassa, Ainaro Subdistrict, Ainaro District, East Timor; Right: Man in Fatuc Laran, Lactos, Cova Lima District, East Timor, 2009. Source: https://en.wikipedia.org/wiki/Bunak_people ..... - 78 -
Figure 124 Sacred house (Lee Teinu) in Lospalos ..... - 78 -
Figure 125 Wooden houses in an old Chinese village in Yunnan, China ..... - 79 -
Figure 126 Vernacular architecture with a typical Chinese courtyard; bottom photos are AHJ with local folk in a Muslim restaurant in Kunming City, Yunan ..... - 79 -
Figure 127 Dai Theravada Buddhist temple in Menghai County, Xishuangbanna. https://en.wikipedia.org/wiki/Yunnan#/media/File:Weihan_Manduan_Temple_Menghai.jpg ..... - 80 -
Figure 128 Tuogu Mosque in Ludian County. https://en.wikipedia.org/wiki/Yunnan#/media/File:%E6%8B%96%E5%A7%91%E6%B8%85%E7%9C%9F%E5%AF%BA_-_panoramio_-_hilloo_(50).jpg ..... - 80 -
Figure 129 Traditional Dai village house in Jinghong, Yunnan A traditional house in a Dai village near Jinghong, with living space on the top, and storage space below. ..... - 80 -
Figure 130 Traditional Dai village house in Jinghong, Yunnan Traditional, rustic, wooden hut house in a rural village with living space on the top, and storage & work space below ..... - 80 -
Figure 131 Manyangguang Dai Ethnic Village in Jinghong City, XishuangBanna ..... - 80 -
Figure 132 Replica designs in an effort to preserve the Malay culture and architecture in Peninsula Malaysia - 81 -
Figure 133 Wooden Houses with cane roofs at Hainan Island — Photo by ADavydova ..... - 82 -
Figure 134 Auxiliary pile-dwelling of Hainan Li Nationality ..... - 82 -
Figure 135 Hainan's Maona Village promotes rural tourism to increase locals' income and expedite rural revitalization Source: XinhuaEditor: huaxia2022-04-12 21:58:47 ..... - 82 -
Figure 136 Left: Lepidotrigona ventralis; Right: Apis hainanensis ..... - 83 -
Figure 137 Blue-tailed bee-eaters ..... - 83 -
Figure 138 Right: Taiwanese meliponiculture of Lepidotrigona ventralis hoozana; Left: Nest entrance, ... - 84 -
Figure 139 Lepidotrigona ventralis hoozana ..... - 85 -
Figure 140 A stilt house of the Truku people. (Author unknown, Public Domain Creative commons https://en.wikipedia.org/wiki/Architecture_of_Taiwan#/media/File:Mokka_and_their_house.jpg ..... - 86 -
Figure 141 An observational painting of a stilt house of one of the plains indigenous peoples as depicted in Liu Shi Qi's Taiwan Panorama Prints (六十七兩采風圖合卷). ..... - 86 -
Figure 142 Typical Malay Rural Village House 1960s ..... - 86 -
Figure 143 A bee house built in Prawita Garden Apiary in Banyumas, Java, Indonesia ..... - 86 -
Figure 144 Left: View of a Tao tribal Village; Middle: The Kakita'an ancestral house of the Amis people in the Tafalong Tribe (太巴塱部落); Right Stiled-round-ended house of the Puyuma people ..... - 87 -

Figure 145 Left: Traditional bamboo house with a low-pitched roof of the Atayal people; Middle: Thatched roof Houses of the Seediq people in Paran Tribe (巴蘭部落); Right: Conical Houses of the Bunun people - 87 -

Figure 146 Top Left: An imitation animal bone hut of the Tsou people; Top Middle: An imitation granary of the Rukai people; Top Right: An imitation of a ceremonial rack of skulls of the Paiwan people in Formosan Aboriginal Culture Village; Bottom Left: Totem Poles; Bottom Right: The Naruwan Theatre the in Formosan Aboriginal Culture Village .... - 87 -

Figure 147 Earthquake Data 25 Feb to 5Mar 2022 in Sumatra .... - 92 -

Figure 148 Recent earthquake in W. Java 22 Nov. 2022 .... - 93 -

Figure 149 Recent earthquakes in Tual, Maluku Islands as of 22 Nov. 2022 data. .... - 94 -

Figure 150 Traditional Musalaki House .... - 94 -

Figure 151 Traditional Nias Island House .... - 96 -

Figure 152 Proposed Hive Placement Pedestal in flood-prone areas .... - 97 -

Figure 153 Inspired by Langkawi coastal house on stilts. .... - 98 -

Figure 154 Inspired by Qolsharif Mosque in Kazan Kremlin, Tatarstan, Russia .... - 98 -

Figure 155 Floating on floodwaters .... - 99 -

Figure 156 Wooden A-Frame Tree House Bee Shed on stilts .... - 99 -

Figure 157 Inspired by the Korowai Treehouse .... - 99 -

Figure 158 Vernacular Fusion – Tanimbar Is. Dwelling & Tambi House of C. Sulawesi .... - 100 -

Figure 159 Conical roofs in Indonesia .... - 103 -

Figure 160 Selected global conical structures. .... - 104 -

Figure 161 Dome roof structure was underway construction Ilog Maria Honeybee Farm .... - 104 -

Figure 162 Conical roofs in other regions around the world. .... - 105 -

Figure 163 Conical structures in Africa and other regions .... - 106 -

Figure 164 Inspired by a Chiang Rai Bamboo village hut. .... - 107 -

Figure 165 Inspired by the Typical Nepali "Goal Ghar" or roundhouse .... - 107 -

Figure 166 Log hive with a conical roof. .... - 108 -

Figure 167 Stingless bee keeping in South America .... - 108 -

Figure 168 Honai Hut of the Dani Tribe in Papua .... - 108 -

Figure 169 Ebai hut for women of the Dani Tribe .... - 109 -

Figure 170 Waimai hut for the pets, poultry and animals .... - 109 -

Figure 171 Kariwari for children and communal events. .... - 109 -

Figure 172 3- tiered conical roof series .... - 110 -

Figure 173 Dried Gourds, Melons, Squash, Pumpkins and Maja .... - 111 -

Figure 174 Coconut kernels .... - 111 -

Figure 175 Coconut shell halves - Bao Technique .... - 111 -

Figure 176 Bamboo Hives .... - 111 -

Figure 177 Branch hollow Beehives .... - 112 -

Figure 178 Wharf or Jetty floats beehive. .... - 112 -

Figure 179 PVC Pipe Hive .... - 112 -

Figure 180 Typical Sagada clay pot with ears for hanging .... - 113 -

Figure 181 Log Hive .... - 113 -

Figure 182 The wooden Box hive .... - 113 -

Figure 183 Hanging coffins are one of the cultures of Sagada, Mountain Province, Luzon. .... - 113 -

Figure 184 Funerary practices of the Toraja people at the Londa Nanggala Cave in Sulawesi .... - 114 -

Figure 185 Famous hanging coffins are those which were made by the Bo people (now extinct) of Sichuan and Yunnan .... - 114 -

Figure 186 Moveable flower cart that incorporates a Bee shack on a Bullock cart. .... - 115 -

Figure 187 Inspired by a Mongolian Yurt on a Nomadic Trailer .... - 115 -

Figure 188 Moveable Bee House on a trailer – Inspired by a Gypsy Caravan. .... - 115 -
Figure 189 Artist's Bee Shed on a trailer – Inspired by a roadshow buggy. .... - 115 -
Figure 190 Inspired by the original bonnet roof and split into three levels for the bees to forage independently on each level. .... - 116 -
Figure 191 Initially inspired by the Mabuhay house in Samar Island of the Philippines and seems to have similarities to Viking Norse houses. .... - 117 -
Figure 192 The 'Hobbits' and some Vikings had a similar cool idea. .... - 117 -
Figure 193 A Modern Turf roof in Norway, a testament to the Nordic Viking heritage. .... - 118 -
Figure 194 The 'Hobbit' house in Beengo Farm in Leyte, Eastern Visayas. .... - 118 -
Figure 195 Gary built a traditional bee house by the creek that runs by his farm. .... - 118 -
Figure 196 Mabuhay house on Samar Island and a Green roof bee shack. .... - 119 -
Figure 197 Inspired by the traditional Mabuhay house of Samar Is. .... - 119 -
Figure 198 Green Roofed Boathouse model examples (excerpts from https://medium.com/goplaceit-datos/fant%C3%A1sticas-casas-flotantes-e9d697caaac8 .... - 120 -
Figure 199 Some of the Mosques in Thailand. Images from Wikimedia .... - 121 -
Figure 200 Inspired by the King Faisal Mosque in Islamabad, Pakistan. .... - 121 -
Figure 201 On the left are Calligraphy and Islamic designs in Thailand. A mosque-inspired bee housing on the right in north Peninsula Malaysia. .... - 122 -
Figure 202 Left: Tranquerah Mosque in Malacca; Right: Proposed tiered three roof bee house/ .... - 122 -
Figure 203 Pyramidal top mosques in Melaka and Riau. .... - 122 -
Figure 204 More examples of three-tiered roofing of mosques in Sulawesi and Lampung .... - 123 -
Figure 205 Sheikh Karimul Makhdum Mosque. Tawi Tawi Island .... - 123 -
Figure 206 Ziggurat Bee House, a multi-tiered Green Roof structure inspired by "The Hanging Gardens of Babylon". .... - 124 -
Figure 207 Green roof bee house inspired by pyramidal tired roof mosques. .... - 124 -
Figure 208 This hand-coloured engraving, probably made in the 19th century after the first excavations in the Assyrian capitals, depicts the fabled Hanging Gardens, with the Tower of Babel in the background. Public Domain File: Hanging Gardens of Babylon. Jpg Created: 19th-century; Unknown author - http://www.plinia.net/wonders/gardens/hgpix1.html .... - 125 -
Figure 209 Traditional house with balconies and a veranda in Negeri Sembilan, Malaysia .... - 125 -
Figure 210 Wind Breakers and Effects .... - 126 -
Figure 211 PROTECTION FROM GALES .... - 128 -
Figure 213 PROTECTION FROM STRONG WINDS .... - 128 -
Figure 214 Ventilation by Natural Elements on Octagon Structure - Effects of Beescape on internal air circulation .... - 130 -
Figure 215 Air circulation and Beescape for Ibaloi House .... - 130 -

List of Maps

Map 1 Aru Islands .... - 57 -
Map 2 Map of Melanesia .... - 65 -
Map 3 Arafura Sea Edited by M. Minderhoud - https://ms.wikipedia.org/wiki/Fail:Locatie_Arafurazee.PNG .. - 73 -
Map 4 Vernacular architecture in Timor Leste, West Timor & Rote Island, NTT, Indonesia. .... - 75 -
Map 5 Migration pattern showing the origin of Malays according to developed theories. .... - 81 -

Epilogue

Growing old gracefully is one of the sternest tests that life sets us. Whether you are a building being restored or a human being thinking of a facelift, it is important to be good at ageing.

We live in a world that admires the new, the instant, and the young and has little time for the old, but art has always known that time is a friend, not an enemy. Old things have a beauty that new things can never have. It's a beauty that's been earned. Waldemar Januszczak

The old indigenous ways are a culmination of years of ancient knowledge. When nurtured in a rural scenario, one's appreciation of the vernacular ambience shapes and moulds one's outlook on ancient human dwellings free from colonial influences. Modernization is not to be ruled out in the progress of civilization. But suppose one is to consider nature and its components. In that case, one cannot overlook the importance of nature's tiny members that keep everything alive.

These tiny members, especially some bees, may be sharing your dream home, and you feel it inconvenient; you must believe that you have built a home so grand and worthy of the greatest tiny being that ever existed on this earth. If your faith does not agree with the concept that God inspired the bees, you need not destroy their nest or chase them away. Instead, build another grand home for them. Hopefully, this book has inspired you to understand their (the bees) needs and the importance of their comfortable existence in your pursuit of health, wealth and happiness.

So much we can learn from these tiny bees if only we take the time and effort to care for and house them appropriately. Their presence is not a nuisance; they may bring unlimited fortunes.

~ APPENDICES ~

Appendix A - Glossary of Roofing Terms and Definitions

A compilation of terms and their definitions that apply to roofing from various dictionaries, including https://www.owenscorning.com/en-us/roofing/tools/roofing-glossary. This list includes applicable terms related to guttering as it pertains to roofing.

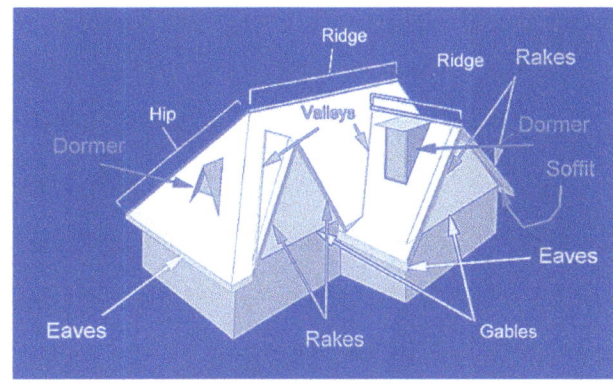

Image 1 Diagram of the main architectural elements that make up a pitched roof.

A

Asphalt: A bituminous waterproofing agent applied to roof materials during manufacturing.

Attic: The open area above the ceiling and under the roof deck of a steep-sloped roof.

Attic Vent: An opening that releases air, heat, and water vapour from the attic. This vent can help prevent damage to the roofing materials caused by overheating.

B

Bargeboard: A **rake fascia** is a board fastened to each projecting gable of a roof to give it strength and protection and to conceal the otherwise exposed end grain of the horizontal timbers or purlins of the roof.

Base Ply: The primary ply of roofing material in a roofing system.

Base Sheet: An asphalt-impregnated or coated felt is used as the first ply in some built-up and modified bitumen roof systems.

Battens: 1" x 2" x 4' wood strips nailed to the roof, upon which a field tile. The metal panels are sometimes hung on battens, but the battens for metal roofing panels are typically much larger.

Bird Stop: A clay or metal product used at the eave of a profile roof to stop birds from entering below the tile.

Blisters: Bubbles that may appear on the surface of asphalt roofing after installation.

Blind Nailing: Nails driven in such a that the succeeding layer of roofing conceals the nail heads themselves.

Bridging: A method of reroofing with metric-sized shingles.

Broken tiles: tiles installed on a roof surface that have broken to where they need to be replaced.

Bonnet Roof: A type of hip roof, often with a sharp slope at the peak and a gentle slope at the base, but its defining characteristic is that it extends beyond the vertical walls, creating a covered space for porches or patios, much like a bonnet.

Booster Tile: Normally 3"-4" long tile strip is used to lift the cover tile. Sometimes, it is used to boost field tiles to create an authentic-looking roof.

Built-up roof: A flat or low-sloped roof consisting of multiple layers of asphalt and ply sheets. Also known as BUR (Built-Up Roofing)

Butt edge: The lower edge of the shingle tabs.

C

Cant Strip: A bevelled piece of wood, fibreboard, or metal strips used at the junction in the roof where the deck meets a vertical wall used to break a hard right angle. Primarily used in low-sloped roofing to prevent the normal cracking of roofing materials applied up a vertical wall at a 90-degree angle.

Cap Flashing: Flashing installed on top of a parapet wall prevents water flow from getting behind the roofing installed on the parapet wall.

Caulk: To fill a joint with mastic or asphalt cement to prevent leaks.

Cathedral Ceiling: a ceiling that follows the angle of the roof up to the highest point towards to top of the roof. It has straight edges (not curved), unlike a vaulted ceiling. This feature in a room will make the room appear bigger and more open.

Clay tile: Tile is made from mostly clay instead of concrete or slate tile, which is not made from clay. Clay tile is typically referred to as "Spanish Tile".

Coating: A layer of viscous asphalt applied to the base material into which granules or other surfacing is embedded.

Collar: Pre-formed flange was placed over a vent pipe to seal the roof around the vent pipe opening. Also called a vent sleeve.

Concealed nail method: Application of roll roofing in which all nails are driven into the underlying course of roofing and covered by a cemented, overlapping course. Nails are not exposed to the weather. AKA Blind nailing.

Condensation: The change of water from vapour to liquid when warm, moisture-laden air comes in contact with a cold surface.

Concrete tile: Roofing tile is made from concrete instead of clay or slate. Also sometimes known as ceramic tile.

Cornice: A cornice (from the Italian cornice meaning "ledge") is generally any horizontal decorative moulding that crowns a building or furniture element—for example, the cornice over a door or window, around the top edge of a pedestal or along the top of an interior wall.

Counter flashing: That portion of the flashing is attached to a vertical surface to prevent water from migrating behind the base flashing.

Counter Battens: Vertical furring strips running beneath and perpendicular to the horizontal tile batten allow drainage and airflow beneath the roof tile. Also known as strapping.

Cricket: A peaked saddle construction at the back of a chimney to prevent snow and ice accumulation and deflect water around the chimney.

Crow's Nest: The term is sometimes used metaphorically for the topmost structures in buildings or towers. A crow's nest provides a good lookout point (hence the name) when built in an elevated position on buildings.

D

Damper: An adjustable plate for controlling the draft.

Dead / Dead Level: Refers to a deck or roof that is not pitched at all.

Dead Load: A non-moving (static) rooftop load, such as mechanical equipment, air-conditioning units, and the roof deck.

Deck: The surface installed over the supporting framing members to which the roofing is applied.

Dormer: A framed window unit projecting through the sloping plane of a roof.

Downspout: A pipe for draining water from roof gutters. Also called a leader.

Drip edge: A non-corrosive, non-staining material used along the eaves and rakes to allow water run-off to drip clear of underlying construction.

Dry Rot: Fungal timber decay occurs in poorly ventilated buildings, resulting in cracking and powdering of the wood.

Dutch lap method: Application of giant individual shingles with a long dimension parallel to the eaves. Shingles are applied to overlap adjacent shingles in each course and the course below.

E

Eaves: The horizontal, lower edge of a sloped roof.

Eave Board: Wood boards run along the roof edge and extend beyond the home's exterior walls.

Eaves flashing: An additional layer of roofing material is applied at the eaves to help prevent damage from water backup.

Eave Riser: A piece of metal used to elevate the starter course of the tile to the appropriate height. It should have pre-drilled weep holes to let the water that naturally gets under most tile roofs drain out from under it, hopefully into a gutter.

Edging strips: Boards nailed along eaves and rakes after cutting back existing wood shingles to provide secure edges for reroofing with asphalt shingles.

Ell: An extension of a building at right angles to its length.

F

Fascia: A decorative board that conceals the rafters' lower ends or the roof's outer sides. Fascia is usually necessary for installing gutters.

Felt: Fibrous material saturated with asphalt and used as an underlayment or sheathing paper.

Fibreglass mat: An asphalt roofing base material manufactured from glass fibres.

Finial: A distinctive section or ornament at the apex of a roof, canopy, etc., on a building.

Flange is a metal pan extending up or down a roof slope around flashing pieces.

Flashing: Pieces of metal or roll roofing are used to prevent water seepage into a building around any intersection or projection in a roof, such as vent pipes, chimneys, adjoining walls, dormers and valleys. Galvanized metal flashing should be a minimum of 26 gauge.

Flashing Cement: See asphalt plastic roofing cement.

Flat Tile: Concrete tile made in the flat or "shake" looking profile. It is without curves and very flat looking.

G

Gable: The upper portion of a sidewall that comes to a triangular point at the ridge of a sloping roof.

Gable roof: A type of roof containing sloping planes of the same pitch on each ridge side. Contains a gable at each end.

Gambrel roof contains two sloping planes of different pitches on each ridge side. The lower plane has a steeper slope than the upper. Contains a gable at each end.

Garret: A garret is a habitable attic, a living space at the top of a house or larger residential building, traditionally small, dismal, and cramped, with sloping ceilings.

Gutter: The trough that channels water from the eaves to the downspouts.

Gutter apron – A metal strip that goes under the shingles, over your eave, or into the gutter.

H

Hip: The inclined external angle formed by the intersection of two sloping roof planes. Runs from the ridge to the eaves.

Hip roof: A type of roof containing sloping planes of the same pitch on each of the four sides. Contains no gables.

Hip shingles cover the inclined external angle formed by the intersection of two sloping roof planes.

L

Lap: To cover the surface of one shingle or roll with another.

Lap cement: An asphalt-based cement used to adhere to overlapping plies of roll roofing.

Lightweight Roofing Tile: Roof tile of a mass/unit area weight of less than 9 lbs per square foot of installed weight, excluding all other roofing components.

Louvre: A slanted opening for ventilation.

M

Mansard roof: A type of roof containing two sloping planes of different pitches on each of the four sides. The lower plane has a much steeper pitch than the upper, often approaching vertically. Contains no gables.

Modified Bitumen: Roll roofing membrane made with polymer-modified asphalt.

Monitor: A monitor in architecture is a raised structure running along the ridge of a double-pitched roof, with its roof running parallel to the main roof.

Mortar: Concrete mixed and made on the job site with tile roofs. Used very similarly as a grout for a tile floor.

N

Nailer Board/Stringer: A piece of wood or other material of proper height and length attached to a roof at a hip or ridge intersection used to allow for proper support and means of attachment for hip and ridge tiles. They can also be used in a pan and cover tile for proper support (also known as a vertical stringer)

Natural ventilation: A ventilation system utilizing ventilators installed in openings in the attic and properly positioned to take advantage of natural airflow to draw hot summer or moist winter air out and replace it with fresh outside air.

Nesting: A method of reroofing with new asphalt shingles over old shingles in which the top edge of the new shingle is butted against the bottom edge of the existing shingle tab. Also, see bridging.

Normal slope application: Method of installing asphalt shingles on roof slopes between 4 inches and 21 inches per foot.

O

Overhang: That portion of the roof structure that extends beyond the exterior walls of a building.

Overlapping Gable: A roof with a double gable where one overlaps another with a slight overhang.

P

Pan Tile: This refers to a piece of tile used in a 2-piece tile roof system that consists of tops and pans. Pantiles catch and direct the water off the roof, and the top tiles shed the water onto the pan tiles. These tiles are typically tapered.

Parapet: A low protective wall along the edge of a roof, bridge, or balcony.

Pediment: The triangular upper part of the front of a classical building, typically surmounting a portico.

Pitch: The degree of roof incline is expressed as the ratio of the rise, in feet, to the span.

Ply: The number of roofing layers: i.e., one-ply, two-ply.

Purlin: A horizontal beam along the length of a roof, resting on principals and supporting the common rafters or boards.

R

Racking: Roofing application method in which shingle courses are applied vertically up the roof rather than across and up. It's not a recommended procedure.

Rafter: The supporting framing member is immediately beneath the deck, sloping from the ridge to the wall plate.

Rake: The inclined outer edge of a sloped roof that extends over a wall from the eave up to the ridge.

Rake board: It gives your roof and eaves a finished look you want to enhance the ambience of the exterior appearance.

Ridge: The uppermost horizontal external angle formed by the intersection of two sloping roof planes.

Ridge shingles cover the horizontal external angle formed by the intersection of two sloping roof planes.

Ridge Vent: An air slot cut into the roof deck at the highest peak on the roof.

Rise: The vertical distance from the eaves line to the ridge.

Roll roofing: Asphalt roofing products are manufactured in roll form.

Roof ornament: In architecture, it is a small decorative device employed to emphasize the apex of a dome, spire, tower, roof, gable or any of various distinctive ornaments at the top, end, or corner of a building or structure.

Run: The horizontal distance from the eaves to a point directly under the ridge. One-half the span.

S

Saddle Roof: A roof is a roof form that follows a convex curve about one axis and a concave curve about the other. The hyperbolic paraboloid form has been used for roofs since it is easily constructed from straight sections of lumber, steel, or other conventional materials. The term is used because the form resembles the shape of a saddle.

Shed roof: A roof containing only one sloping plane. Has no hips, ridges, valleys or gables.

Skylight: A window set in a roof or ceiling at the same angle.

Slope: The degree of roof incline is expressed as the ratio of the rise, in inches, to the run, in feet.

Soffit: The finished underside of the eaves.

Soffit Vent: An under-eave opening is needed to intake outside air into an attic space.

Span: The horizontal distance from eaves to eaves.

Spire: A tapering conical or pyramidal structure on the top of a building.

Strut: A rod or bar forming part of a framework to resist compression.

T

Tab: The exposed portion of strip shingles is defined by cutouts.

Thermal Sealant: Sealant that is heat activated. Usually applied by the manufacturer to the back of a shingle. Once the heat is activated at about 70 degrees, it bonds the shingles together.

Tile Pan Metal: This is an accessory piece used where a tile roof terminates into a vertical sidewall. The metal catches the water run-off from the vertical and keeps it from getting under the tile roof and eventually directs the water back onto the top of the tile roof even though it is installed under the tile. It has different widths; a 6" width is better than a 4" tile pan.

T-Lock Shingles: These were popular in the 1930s and regularly used on houses through the 1980s. The shingles have a T-shaped design that allows them to interlock.

Top lap: The succeeding course after installation covers that portion of the roofing.

Tongue and Groove: wooden planking in which adjacent boards are joined utilizing interlocking ridges and grooves down their sides. Also known as T&G.

Truss: A framework, typically consisting of rafters, posts, and struts, supporting a roof, bridge, or other structure.

V

Valley: The internal angle formed by the intersection of two sloping roof planes to provide water runoff.

Vaulted ceiling: A ceiling curved up towards the top of the highest point towards the top of the roof. This feature in a room will make the room appear bigger and more open.

Vent: Any air outlet that protrudes through the roof deck, such as a pipe or stack. Any device installed on the roof, gable or soffit to ventilate the underside of the roof deck.

Vent sleeve: See collar.

Ventilators: Devices that eject stale air and circulate fresh air (i.e., ridge, roof, gable, under the eave, foundation or rafter vents and vented soffit panels.)

W

Weathervane: A Revolving pointer to show the direction of the wind, typically mounted on top of a building.

Appendix B – List of roof types

- **List of roof types**
  - 1. A-Frame Roof
  - 2. Barrel Vaulted Roof
  - 3. Bell Roof
  - 4. Box Gable Roof
  - 5. Butterfly Roof
  - 6. Clerestory Roof
  - 7. Combination Roof
  - 8. Conical Roof
  - 9. Cross-Hipped Roof
  - 10. Curved Roof
  - 11. Dome Roof
  - 12. Domed Vault Roof
  - 13. Dormer Roof
  - 14. Dropped Eaves Roof
  - 15. Dutch Gable Roof
  - 16. Flat Roof
  - 17. Gable Roof
  - 18. Gambrel Roof
  - 19. Hexagonal Roof
  - 20. Hip and Valley Roof
  - 21. Hip Roof
  - 22. Jerkinhead Roof
  - 23. M-Shaped Roof
  - 24. Mansard Roof
  - 25. Monitor Roof
  - 26. Pyramid Hip Roof
  - 27. Saltbox Roof
  - 28. Sawtooth Roof
  - 29. Shed Roof
  - 30. Skillion Roofing

Index

A

Aborlan, - 13 -
acute cone, - 103 -
Adjacent Hillock, - 129 -
adornments, - 122 -
aesthetic appeal, - 16 -
Aeta's, - 24 -
A-Frame, - 147 -
Aguaruna, - 13 -
air mattress, - 99 -
air movement, - 60 -
Amazon, - 97 -
Ampang, - 97 -
Anacardiaceae, - 84 -
anahaw, - 24 -
animal skin, - 103 -
ankaw, - 31 -
ants, - 113 -
Apokon Road, - 49 -
arabesque, - 59 -
aristocrats, - 35 -, - 37 -
Aru Islands, - 57 -
Asmat, - 74 -
Asphalt, - 139 -, - 144 -
Ata Manobo, - 30 -
attic, - 15 -
Attic, - 139 -
Australia, - 59 -
awning, - 125 -

B

Babylon, - 124 -
Babylonia, - 124 -
Bagani, - 27 -, - 30 -
Baghdad, - 124 -
Bagobo Tagabawa, - 32 -
Bahay Kubo., - 42 -
Bajao, - 27 -
Bajo Cachiaco, - 13 -
Bakag, - 32 -
Bale, - 32 -
Bali, - 112 -
Balinese, - 18 -
Balinese inspired, - 18 -
bamboo groves, - 111 -
Bamboo hives, - 111 -
bamboo lath, - 31 -
bamboo raft, - 99 -
bamboo shingles, - 30 -
bamboo slats, - 24 -
Banaba, - 65 -
Bao, - 111 -
Bargeboard, - 139 -
bark, - 24 -, - 29 -, - 30 -, - 31 -, - 72 -, - 104 -
Barrel Vaulted, - 147 -
basement, - 110 -
Batanes Island, - 19 -
Battens, - 139 -, - 141 -
Baylan, - 29 -
beasts of burden, - 115 -
Bee repositories, - 98 -
bee sheds, - 14 -
Bee sheds, - 117 -
Bee-eaters, - 83 -
Beekeepers associations, - 98 -
Beengo Farm, - 118 -
Bees *for* Development, - 12 -
Bell Roof, - 147 -
benefit over foraging space cost, - 117 -
Benguet, - 44 -
Benteng None, - 75 -
berries, - 84 -
Bevak, - 72 -
Bicol, - 111 -
Bicol Region, - 47 -
Binu PT, - 112 -
Bird Stop, - 139 -
bisj, - 74 -
B'Laan, - 33 -
black netting, - 121 -
Blueprint, - 97 -
boat builders, - 21 -
Bonnet Roof, - 140 -
Bontok, - 23 -
Borneo, - 24 -, - 35 -, - 82 -, - 111 -, - 113 -
Boulder Soil Mound, - 129 -
Box Gable Roof, - 147 -
box hives, - 18 -, - 113 -
box hives rack, - 18 -
Branch hollows, - 112 -
bullock carts, - 115 -
Bure, - 66 -
Butterfly roof, - 121 -
Butterfly Roof, - 147 -

C

Calligraphy, - 122 -
campers, - 115 -
Canberra, - 26 -
canoes, - 97 -
canopy, - 125 -, - 142 -
caravans, - 115 -
carved wings, - 26 -
catwalks, - 27 -
Centralised harvesting, - 117 -
cerana, - 84 -
Chaiyi, - 85 -
Chiang Rai, - 107 -
Chiayi University, - 84 -
China, - 114 -
Clark Kent, - 59 -
Clerestory, - 147 -
climbers, - 125 -
coastal, - 28 -
Cockington Green Gardens, - 26 -
Coconut Kernel, - 111 -
Coconut shell, - 47 -
Coconut shell halves, - 111 -
cogon, - 19 -, - 20 -, - 23 -, - 24 -, - 28 -, - 30 -, - 31 -, - 32 -, - 103 -
cogon grass, - 20 -, - 103 -
Collar, - 140 -
Combination Roof, - 147 -
Compositae, - 84 -
Conical, - 15 -, - 102 -, - 108 -
Conical Roof, - 147 -
conical roofs, - 15 -, - 103 -
Conical roofs, - 102 -
consecutive tremors, - 92 -
Consolidation of safety, - 117 -
construction durability, - 20 -
contaminated hives, - 117 -
Cooperatives, - 98 -
Cornice, - 141 -
Cotabato, - 28 -, - 33 -
cottage designs, - 59 -
crawling enemies, - 113 -
creepers, - 119 -, - 125 -
creepers and climbers, - 119 -
Cricket, - 141 -
Cross-Hipped Roof, - 147 -
Crow's Nest, - 141 -
Cruciferae, - 84 -
cucumbers, - 84 -
Cucurbitaceae, - 84 -
Curved Roof, - 147 -

D

Damper, - 141 -
Dani tribe, - 108 -
Danizon farm, - 18 -
Date palm, - 125 -
Datu, - 25 -, - 28 -, - 37 -
Davao City, - 29 -
Davao del Norte, - 29 -, - 30 -

Davao Gulf, - 32 -
Davao Oriental, - 27 -, - 33 -
David Roubik, - 13 -
dietary requirements, - 12 -
Diurnal, - 127 -
dome hemisphere, - 103 -
Dome Roof, - 147 -
Dome Roofs, - 102 -
Domed Vault Roof, - 147 -
dome-shaped, - 102 -
Dormer Roof, - 147 -
Drainage boards, - 116 -
dried grasses, - 12 -
drip irrigation reticulation, - 125 -
Dropped Eaves, - 147 -
Dumaguete, - 111 -
Dutch Gable Roof, - 147 -
Dutch lap, - 141 -

E

Earthquake Dilemma, - 92 -
earthquakes, - 92 -
Eave Board, - 141 -
Ebai, - 109 -
Edhy Yanshah, - 99 -
Ell, - 142 -
Ende, - 94 -, - 95 -
Ende Lio, - 94 -, - 95 -
entrance anchors, - 97 -
estuary, - 74 -
ethnolinguistic, - 19 -
Euphrates River, - 124 -
extreme weather, - 20 -

F

fale, - 20 -, - 21 -
fan-palm leaves, - 111 -
Faraday's cage, - 99 -
Fascia, - 142 -
Fengshan, - 85 -
Fiji Islands, - 66 -
Filipino bee farms, - 16 -
Filipinos, - 19 -
finial, - 121 -
Finial, - 142 -
finials, - 29 -
Fisi, - 67 -
Flat Roof, - 147 -
flat wooden slats, - 103 -
flattened bamboo, - 32 -, - 33 -
flavibasis, - 79 -
flood levels, - 99 -
flooding, - 97 -
flood-prone areas, - 97 -
Flores, - 108 -
Flores Island, - 75 -, - 108 -
flowering ramblers, - 125 -

FOLK Architecture, - 19 -
frogs, - 113 -
frond petiole, - 125 -

G

Gable, - 142 -, - 147 -
gable and hip roofs, - 15 -
Gable Roof, - 147 -
gale, - 19 -, - 91 -, - 127 -
Gambrel roof, - 142 -
Gambrel Roof, - 147 -
garden conservatory, - 116 -
Garret, - 142 -
Garybee, - 118 -
gazebos, - 16 -
geotextile lining, - 116 -
Gono, - 28 -
Gourds, - 111 -
Grajo's Bee Farm, - 16 -
green roof, - 104 -, - 110 -, - 119 -, - 125 -
Green Roofs, - 116 -
Guangxi, - 79 -
Guapito, - 42 -
guava, - 84 -
Gumne, - 33 -
Gunung Mutis, - 75 -
Gusts, - 127 -
Gutter, - 142 -
gypsy vardo, - 115 -

H

Hainan, - 79 -, - 82 -, - 83 -, - 88 -
hainanensis, - 83 -
Halmahera, - 153 -
Hatin, - 82 -
heat dissipation, - 60 -
heavy rains, - 15 -
Hesty, - 72 -
Hexagon Beehouse, - 16 -
Hexagonal, - 16 -
Hexagonal Roof, - 147 -
Hindu, - 23 -
Hip, - 143 -
Hip and Valley Roof, - 147 -
Hip Roof, - 147 -
Hive Design exhibition, - 59 -
Honai, - 108 -, - 109 -
Honai House, - 73 -
hoozana, - 84 -, - 85 -
horse carriage, - 115 -
Hospital, - 117 -
hurricane-proof, - 102 -
hyperbolic, - 103 -

I

Ibaloi, - 21 -
Ifugao, - 20 -
Igorot, - 21 -
igudut, - 21 -
Iloilo, - 41 -, - 43 -
inclined embankment, - 116 -
India, - 104 -, - 112 -, - 113 -
indigenous, - 103 -
INDIGENOUS Architecture, - 19 -
Indo-Malaya, - 113 -, - 121 -
inverted boat, - 21 -
Iraq, - 124 -
Islamabad, - 121 -
Islamic Structures, - 121 -
Isneg, - 21 -
Israel, - 104 -, - 125 -
Ivatan, - 19 -
Ivatans, - 20 -
Iwang, - 33 -

J

James Cook, - 67 -
Javanese Apiary, - 86 -
Jayapura, - 109 -
Jerkinhead, - 15 -, - 147 -
Jun Tabinga, - 13 -

K

Kalinga, - 22 -
Kankanay, - 21 -
Kanume, - 72 -, - 104 -
Kaohsiung, - 85 -
Kariwari, - 109 -
kawit, - 29 -
Kelantanese, - 80 -
Kerala, - 112 -
King Faisal Mosque, - 121 -
Kiribati, - 65 -
Kaohsiung, - 85 -
Korowai, - 99 -, - 108 -
Kuala Lumpur, - 97 -

L

La Hisa, - 72 -
Lake Lahit, - 28 -
Lake Sebu, - 28 -
Lake Selutan, - 28 -
Lake Sentani, - 109 -
Lanao, - 25 -
Landslide, - 97 -
Langkawi, - 80 -, - 97 -
lapong, - 32 -
lawaan, - 29 -, - 30 -, - 31 -
Leo Grajo, - 16 -

Lepidotrigona, - 85 -
Lepidotrigona ventralis, - 85 -
Lepidotrigona ventralis hoozana, - 84 -, - 85 -
limestone walls, - 20 -
linapakan, - 32 -
Lio, - 94 -
lithographs, - 66 -
livestock, - 109 -
lizards, - 113 -
Load bearing struts, - 117 -
Log hives, - 113 -
Lombok, - 111 -
Louvre, - 143 -
Lumawig, - 23 -
Luzon, - 19 -
lyupakan, - 34 -

M

Mabuhay house, - 119 -
Maharlika, - 35 -
Maja, - 111 -
Malaysia, - 97 -, - 113 -, - 122 -
Mandaya, - 27 -, - 29 -, - 33 -
Mangguangan, - 29 -
mango, - 84 -
Mangyan, - 24 -
Manobo, - 29 -, - 30 -, - 32 -
Manokwari, - 73 -
Mansaka, - 29 -
Mansard roof, - 143 -
Mansard Roof, - 147 -
Maranao, - 25 -, - 26 -
Marble, - 104 -
Marinduque, - 111 -
Matikadong, - 29 -
Mbaru Niang, - 108 -
Mbus Tree, - 72 -
Melaka, - 122 -
Melaleuca, - 72 -, - 104 -
Melaleuca cajuputi, - 72 -
Melaleuca leucadendra, - 72 -
Melanesia, - 65 -, - 66 -, - 69 -
melon, - 84 -
Merauke, - 72 -
Mesopotamia, - 124 -
Mesozoic, - 82 -
MGVI, - 97 -
minaret, - 121 -
Mindanao, - 19 -, - 25 -, - 28 -, - 29 -
Mindoro, - 18 -, - 111 -
miniatures, - 26 -
Mobile Racks, - 115 -
Monitor, - 143 -
Monitor Roof, - 147 -
Mosques, - 121 -, - 122 -
motorised equipment, - 115 -

Mount Apo, - 32 -
M-Shaped Roof, - 147 -
multi-tiered, - 125 -
Musalaki, - 94 -, - 95 -
Muslim Beekeepers, - 121 -
Myrtaceae, - 84 -

N

Nantou, - 85 -
National Park, - 72 -
natural fibres, - 103 -
Negros, - 111 -
Nesting, - 143 -
Nias Island, - 96 -
Noche, - 19 -
nomadic tents, - 115 -
Nora Peña, - 13 -
Nordic Viking, - 118 -
Norway, - 118 -
notch ladder, - 99 -
NTB, - 111 -
Nusa Tenggara, - 75 -, - 91 -, - 94 -, - 95 -
Nusantara, - 123 -

O

Octagonal House, - 22 -
oilseed rape, - 84 -
old cottage, - 59 -
Omo Hada, - 96 -
Orchidaceae, - 84 -
Overhang, - 144 -
Overlapping Gable, - 144 -

P

Padilla, - 83 -, - 88 -
Pakistan, - 121 -
Palawan, - 13 -
palm fronds, - 103 -
palm leaves, - 12 -, - 13 -
papaudan, - 30 -
paperbark, - 72 -, - 104 -
papo ua, - 72 -
Papua, - 72 -, - 104 -, - 108 -, - 109 -
parabolic, - 103 -
pear, - 84 -
Pediment, - 144 -
pediments, - 121 -
pepuah, - 72 -
perpetual conflict, - 96 -
Perspex, - 121 -
Pesta Air, - 97 -
Philippines, - 1 -, - 13 -, - 15 -, - 19 -, - 20 -, - 47 -, - 93 -, - 111 -, - 118 -, - 119 -, - 123 -
Phoenix sp, - 125 -

Pia's Bee Farm, - 48 -
Pilar, - 47 -
Pile-dwelling, - 88 -
Pitch, - 144 -
platform, - 97 -
plum, - 84 -
pollination program, - 117 -
pollination services, - 115 -
porch, - 125 -
Portuguese, - 36 -
precise technical application, - 117 -
Pre-colonial architecture, - 19 -
Pre-colonial Housing, - 19 -
prehistoric stone age, - 99 -
prevailing wind, - 127 -
PRIMITIVE Architecture, - 19 -
Puerto Princesa, - 14 -
Pumpkins, - 111 -
Purlin, - 144 -
PVC pipe hives, - 112 -
Pyramid Hip, - 147 -
pyramidal, - 122 -

Q

Quarantine, - 117 -

R

Rafter, - 144 -
Raja Ampat, - 99 -
Rake, - 144 -
rat guard, - 31 -
rattan leaves, - 29 -, - 32 -
reeds, - 103 -
replica, - 108 -
Repository, - 97 -, - 110 -
research stations, - 98 -
rice stalks, - 103 -
Ridge, - 144 -
Rise, - 144 -
roadshow buggy, - 115 -
Rock outcrops, - 129 -
Romblon, - 42 -
Roof ornament, - 145 -
rooftops, - 116 -
Rosaceae, - 84 -
Roubik, - 13 -
Round Barn, - 110 -
round houses, - 46 -, - 102 -
royal nobility, - 35 -
Run, - 145 -

S

Saddle Roof, - 145 -
Sagada, - 44 -, - 113 -
salahiya, - 32 -
Saltbox roof, - 15 -

Saltbox Roof, - 147 -
Samar, - 111 -, - 119 -
Samar Island, - 119 -
Sanskrit, - 35 -
Sapindaceae, - 84 -
Sarimanok, - 26 -
sasal, - 30 -
Sawtooth Roof, - 147 -
Sekayu, - 97 -
shack, - 116 -
Shed roof, - 145 -
Shed Roof, - 147 -
Silay City, - 111 -
sinasa, - 33 -
sinasa na kawayan, - 32 -
Singapore, - 152 -, - 153 -
single log ladder, - 32 -
Skylight, - 145 -
Slanting Roof, - 129 -
Soffit Vent, - 145 -
solar panels, - 118 -
Solomon Islands, - 65 -
Sorsogon, - 16 -, - 47 -
South China, - 79 -, - 88 -
Spaniards, - 20 -
spear, - 32 -
Spire, - 145 -
split bamboo roof, - 34 -
sponge cucumber, - 84 -
squall, - 127 -
Squash, - 111 -
Sta. Cruz, - 32 -
Stabilized Berms, - 129 -
stairway, - 97 -, - 99 -
steep roof, - 19 -
stilts, - 97 -, - 99 -
stone houses, - 20 -
stone slabs, - 104 -
storage platform, - 21 -
Strand, - 85 -
straws, - 12 -
Strut, - 145 -
submontane, - 85 -
Sulawesi, - 72 -
Sulu, - 123 -
Sumatra, - 92 -
sunflower, - 84 -
Sung, - 84 -
sungag, - 29 -
swamp areas, - 97 -

T

Tagbanwa, - 24 -
Tagum city, - 49 -
Taichung, - 85 -
Taiwan, - 84 -, - 85 -
Taiwan Tribal, - 87 -
T'boli, - 28 -, - 29 -
Telephone booth, - 59 -
Terengganu, - 97 -
terminal passage, - 20 -
Ternate, - 72 -
Tetragonula sapiens, - 65 -
Thailand, - 107 -, - 122 -
thermal comfort, - 60 -, - 117 -, - 129 -
Thermal Sealant, - 145 -
Thick Walls, - 129 -
Tibet, - 79 -
Tidore, - 72 -
Tim Heard, - 59 -
Timor, - 75 -, - 77 -, - 78 -, - 91 -, - 152 -
Timor Leste, - 75 -
Tobati-Enggros tribe, - 109 -
Tog'gan, - 29 -
Tonga, - 67 -
Tongue and Groove, - 145 -
tornado winds, - 102 -
torogan, - 25 -
Torogan, - 26 -
Tortoise, - 12 -
trailers, - 115 -
Tranquerah, - 122 -
Transition bay, - 117 -
tree bark, - 103 -
Tree House, - 99 -
Trigona, - 85 -
Tropics, - 12 -
truncated conical, - 107 -
Truss, - 145 -
Tunga, - 118 -
turf, - 118 -
turf roof, - 118 -
Typhoon prone, - 12 -
Typhoons, - 14 -, - 117 -

U

upturned crescent, - 121 -

V

Valley, - 146 -
valleys, - 29 -, - 32 -, - 108 -, - 142 -, - 145 -
Vaulted ceiling, - 146 -
ventilation, - 23 -, - 143 -
Ventilators, - 146 -
ventralis, - 85 -
Vietnam, - 111 -
vines, - 125 -
Visayas, - 19 -, - 111 -, - 118 -
Viti, - 67 -
Viti Levu, - 67 -

W

Wamai, - 109 -
Wasur, - 72 -
water tanks, - 125 -
watermelon, - 84 -
water-soaked soil, - 117 -
Weathervane, - 146 -
West Papua, - 72 -, - 73 -, - 99 -
Winter melons, - 111 -
withstand strong winds, - 12 -, - 20 -
woven bamboo strips, - 28 -, - 32 -

X

Xinhua News, - 83 -

Y

Yakan, - 27 -
Yanggandur, - 72 -
Yunnan, - 79 -, - 81 -, - 82 -, - 88 -, - 89 -, - 114 -
yurts, - 115 -

Z

Zheng, - 79 -
Ziggurat, - 125 -

Bibliography

Anonymous (n.d.). Indigenous Peoples in the Philippines. *Cordillera Peoples Alliance, Minority Rights Group.*

Ahmed I. & McDonnell T. (2020). Prospects and constraints of post-cyclone housing reconstruction in Vanuatu drawing from the experience of tropical cyclone Harold. Progress in Disaster Science.

Anggraeni, D. (2011). Another East: Representation of Papua in Popular Media. Conference: International Conference on Indonesian Studies: *University of Indonesia, Depok, Indonesia.*

Baja, S. M., & Mostoles, M. D. (2017). Colony Characterization And Amplicon Sequencing Of Stingless Bees (Tetragonula biroi). In A. H. Jalil, *Handbook of Meliponiculture vol 2. Akademi Kelulut Malaysia.*

Birt., M. (n.d.). The Trigona Species (Stingless Bees). Their role in higher plant pollination. *12391833933.*

Cockerell, T. D. (1919). The Social Bees of The Philippines Islands. *Univeraity of Colarado.*

Cortopassi-Laurino, M., & al, e. (2006). Global meliponiculture: challenges and opportunities. *Apidologie 37*, 275–292.

Cossio, B. M. (2005). The Buka–Hatene Community Learning Centre: Friends of Baucau's Project to Rebuild a Community Building in Baucau, Timor Leste. *The University of Chile.*

Couvillon, M. J., et al. (2008). A comparative study in stingless bees (Meliponini) demonstrates that nest entrance size predicts traffic and defensivity. *J . EVOL. BIOL. 21*, 194–201.

Engel, M. S. (2000). A Review of the Indo-Malayan Meliponine Genus Lisotrigona with Two New Species (Hymenoptera, Apidae). *American Museum Of Natural History.*

Hong, N. (2018). Tanimbarese And Rotinese Traditional House. *The National University of Singapore.*

Iyengar, K. (2015). Sustainable Architectural Design. *Routledge.*

Jalil, A. (2014). Beescape for Meliponines. *Singapore: Partridge Publishing.*

Jalil, A. H., & Roubik(ed), D. W. (2019). *Filipino Meliponiculture and Beyond Inc. MIMAROPA Tours*, The Philippines: *Akademu Kelulut Malaysia.*

Jalil, A., & Roubik(ed), D. (2016). Handbook of Meliponiculture. *Akademi Kelulut Malaysia.*

Jalil, A., & Roubik(ed), D. (2018). Handbook of Meliponiculture Vol 2. *Akademi Kelulut malaysia.*

Leonhardt, S. D., & al., e. (2010). Stingless Bees Use Terpenes as Olfactory Cues to Find Resin Sources. *Chem. Senses 35*, 603–611.

Mabahague, E. R. (n.d.). Philippine History Pre Colonial Prtiod. Manila: *University of Santo Tomas.*

Macías-Macías, J. O. et al. (2011). Comparative temperature tolerance in stingless bee species from tropical highlands and lowlands of Mexico and implications for their conservation (Hymenoptera: Apidae: Meliponini). *Apidologie 42*, 679–689.

Madoro G. (n.d.). *Precolonial Housing In The Philippines. Retrieved from Central Colleges of the Philippines*: https://ccp-edu.academia.edu/GlendzMadoro

Michener, C. D. (1961). Observations on the Nests and Behavior of *Trigona* in Australia and New Guinea (Hymenoptera, Apidae). *American Museum Novitates*.

Ntawuzumunsi E., Kumaran S. & Sibomana L. . (2021). Self-Powered Smart Beehive Monitoring and Control System (SBMaCS) †. *Sensors, 21*, 3522.

Occhiuzzi, P. (2000). Stingless bees pollination greenhouse Capsicum. *Aussie Bee 13,15. Published by Australian Nature Bee Research Centre, North Richmond NSW Australia*.

Palittin D. & Hallatu T.G.R. (2019). Sar: Kanume tribal culture in environmental conservation to reduce global warming effects. *IOP Conf. Ser.: Earth Environ. Sci. 235 012062*.

Phoek I. C. A.; Tjilen A. P. & Cahyono E. . (2021). Analysis of Ecotourism, Culture and Local Community Empowerment: Case Study of Wasur National Park - Indonesia. *Macro Management & Public Policies | Volume 03 | Issue 02*.

Rahman1, A., Das, P. K., Rajkumari, P., Saikia, J., & Sharmah, D. (2015). Stingless Bees (Hymenoptera:Apidae: Meliponini): Diversity and Distribution in India. *International Journal of Science and Research (IJSR)*, Volume 4 Issue 1.

Rasmussen. (2008). Catalogue of the Indo-Malayan/Australasian stingless bees (Hymenoptera: Apidae: Meliponini). *Zootaxa*

Rattanawannee A., & Duangphakdee O. (2019). Southeast Asian Meliponiculture for Sustainable Livelihood. DOI: http://dx.doi.org/10.5772/intechopen.90344.

Roubik, D. (2006). Stingless bee nesting biology. *Apidologie 37*.

Sakagami, S. F., & Yamane, S. (1984). Notes on taxonomy and Nest Architecture of the Taiwanese stingless bee *Trigina (Lepidotrigona) ventralis hoozana. Ibaraki University*.

Sakagami, S. F., Inoue, T., Yamane, S., & Salmah, S. (1983). Nest architecture and colony composition of the Sumatran stingless bee *Trigona (Tetragona) laeviceps. Kontyu, 51(1):*, 100-111.

Salatnaya, H. (2019). Potential growth of meliponiculture in West Halmahera, Indonesia. IOP Conference Series *Earth and Environmental Science · December 2019*.

Schwarz, H. F. (1939). The Indo Malayan Species of Trigona. *Bulletin of AMNH*.

Sommeijer, M. (1999). Beekeeping with stingless bees: a new type of hive. *Bee World 80(2):* 70-79.

Starr, C. K., & Sakagami, S. F. (1987). An Extraordinary Concentration of Stingless Bee in the Philippines, with Notes on Nest Architecture (Hymenoptera: Apidae: Trigona spp). *Insectes Sociaux 34(2):96-107*.

Vit, P., Roubik, D. W., & Pedro, S. M. (2013). Pot Honey - A Legacy of Stingless Bees. *Springer, DO - 10.1007/978-1-4614-4960-7*.

Vit, P.; Pedro, S.R.M.; Roubik, D. (2018). Pot-Pollen in Stingless Bee Melittology. *Springer International Publishing AG. DO - 10.1007/978-3-319-61839-5*.

Welzen, P. C., J. F., & Alahuhta, J. (2005). Plant distribution patterns and plate tectonics in Malesia. *Biologiske Skrifter 55, 199-217*.

www.ingramcontent.com/pod-product-compliance
Lightning Source LLC
Chambersburg PA
CBHW061551010526
44117CB00022B/2988